FUNCTIONAL MATERIALS

Properties, Performance, and Evaluation

AAP Research Notes on Chemistry

FUNCTIONAL MATERIALS
Properties, Performance, and Evaluation

Edited by
Ewa Kłodzińska, PhD

A. K. Haghi, PhD, and Gennady E. Zaikov, DSc
Reviewers and Advisory Board Members

Apple Academic Press Inc. | Apple Academic Press Inc.
3333 Mistwell Crescent | 9 Spinnaker Way
Oakville, ON L6L 0A2 | Waretown, NJ 08758
Canada | USA

First issued in paperback 2021

Exclusive worldwide distribution by CRC Press, a member of Taylor & Francis Group
No claim to original U.S. Government works

ISBN 13: 978-1-77463-353-3 (pbk)
ISBN 13: 978-1-77188-037-4 (hbk)

Library and Archives Canada Cataloguing in Publication

Functional materials: properties, performance, and evaluation/edited by Ewa Kłodzińska, PhD; A.K. Haghi, PhD, and Gennady E. Zaikov, DSc, Reviewers and Advisory Board Members.

(AAP research notes on chemistry)
Includes bibliographical references and index.
ISBN 978-1-77188-037-4 (bound)
1. Materials. I. Kłodzińska, Ewa, editor II. Series: AAP research notes on chemistry

TA403.F85 2015   620.1'1   C2015-900261-3

Library of Congress Cataloging-in-Publication Data

Functional Materials (Apple Academic Press)
Functional materials: properties, performance, and evaluation / editor, Ewa Klodzinska, PhD; A. K. Haghi, PhD, and Gennady E. Zaikov, DSc, reviewers and advisory board members.

pages cm. -- (AAP research notes on chemistry)
Includes bibliographical references and index.
ISBN 978-1-77188-037-4 (alk. paper)
1. Materials. I. Klodzinska, Ewa. II. Haghi, A. K. III. Zaikov, G. E. (Gennadii Efremovich), 1935- IV. Title.

TA403.F788 2015   620.1'1--dc23   2014049982

AAP RESEARCH NOTES ON CHEMISTRY

This series reports on research developments and advances in the ever-changing and evolving field of chemistry for academic institutes and industrial sectors interested in advanced research books.

Richard A. Pethrick, PhD, DSc
Research Professor and Professor Emeritus, Department of Pure and Applied Chemistry, University of Strathclyde, Glasgow, Scotland, UK

Charles Wilkie, PhD
Professor, Polymer and Organic Chemistry, Marquette University, Milwaukee, Wisconsin, USA

Georges Geuskens, PhD
Professor Emeritus, Department of Chemistry and Polymers, Universite de Libre de Brussel, Belgium

BOOKS IN THE AAP RESEARCH NOTES ON CHEMISTRY

Chemistry and Chemical Biology: Methodologies and Applications
Editors: Roman Joswik, PhD, and Andrei A. Dalinkevich, DSc
Reviewers and Advisory Board Members: Gennady E. Zaikov, DSc, and
A. K. Haghi, PhD

Functional Materials: Properties, Performance, and Evaluation
Editor: Ewa Kłodzińska, PhD
Reviewers and Advisory Board Members: A. K. Haghi, PhD,
and Gennady E. Zaikov, DSc

High Performance Elastomer Materials: An Engineering Approach
Editors: Dariusz M. Bielinski, DSc, Ryszard Kozlowski, PhD, and
Gennady E. Zaikov, DSc

ABOUT THE EDITOR

Ewa Kłodzińska, PhD

Ewa Kłodzińska holds a PhD from Nicolaus Copernicus University, Faculty of Chemistry in Torun, Poland. For 10 years, she has been doing research on determination and identification of micro-organisms using the electromigration techniques for the purposes of medical diagnosis. Currently, she is working at the Institute for Engineering of Polymer Materials and Dyes and investigates surface characteristics of biodegradable polymer material on the basis of zeta potential measurements. She has written several original articles, monographs, and chapters in books for graduate students and scientists. She has made valuable contributions to the theory and practice of electromigration techniques, chromatography, sample preparation, and application of separation science in pharmaceutical and medical analysis. Dr Ewa Kłodzińska is a member of the editorial boards of *ISRN Analytical Chemistry* and the *International Journal of Chemoinformatics and Chemical Engineering (IJCCE)*.

REVIEWERS AND ADVISORY BOARD MEMBERS

A. K. Haghi, PhD

A. K. Haghi, PhD, holds a BSc in urban and environmental engineering from the University of North Carolina (USA); a MSc in mechanical engineering from North Carolina A&T State University (USA); a DEA in applied mechanics, acoustics, and materials from Université de Technologie de Compiègne (France); and a PhD in engineering sciences from Université de Franche-Comté (France). He is the author and editor of 65 books as well as 1000 published papers in various journals and conference proceedings. Dr. Haghi has received several grants, consulted for a number of major corporations, and is a frequent speaker to national and international audiences. Since 1983, he served as a professor at several universities. He is currently Editor-in-Chief of the *International Journal of Chemoinformatics and Chemical Engineering* and *Polymers Research Journal* and on the editorial boards of many international journals. He is a member of the Canadian Research and Development Center of Sciences and Cultures (CRDCSC), Montreal, Quebec, Canada.

Gennady E. Zaikov, DSc

Gennady E. Zaikov, DSc, is Head of the Polymer Division at the N. M. Emanuel Institute of Biochemical Physics, Russian Academy of Sciences, Moscow, Russia, and Professor at Moscow State Academy of Fine Chemical Technology, Russia, as well as Professor at Kazan National Research Technological University, Kazan, Russia. He is also a prolific author, researcher, and lecturer. He has received several awards for his work, including the Russian Federation Scholarship for Outstanding Scientists. He has been a member of many professional organizations and on the editorial boards of many international science journals.

CONTENTS

LIST OF CONTRIBUTORS

J. N. Aneli
Laboratory for Polymer materials, Institute of Machine Mechanics, Tbilisi 0186, Republic Georgia, Email: jimaneli@yahoo.com

S. Ashimov
Department of Condensed Matter Physics, E. Andronikashvili Institute of Physics, Tbilisi 0177, Republic of Georgia

M. Bazunova
Bashkir State University, Ufa 450076, Republic of Bashkortostan, Russia

S. B. Bokieva
N. N. Semenov Institute of Chemical Physics, RAS, Moscow, Russia

S. N. Bondarenko
Volzhsky Polytechnic Institute, (Branch of Federal State Budgetary Educational Institution of Higher Professional Education) Volgograd State Technical University, Volzhsky 404121, Russia, Email: d.provotorova@gmail.com; www.volpi.ru

V. Buntar
Department of Physics and Astronomy, McMaster University, Hamilton, ON L8S 4M1, Canada

A. E. Chalykh
A. N. Frumkin Institute of Physical Chemistry and Electrochemistry of the Russian Academy of Sciences (IPCE RAS), Moscow, 119991, Russia

J. Chigvinadze
Department of Condensed Matter Physics, E. Andronikashvili Institute of Physics, Tbilisi 0177, Republic of Georgia

A. N. Danilenko
Institute of Biochemical Physics N. M. Emanuel Academy of Sciences, Moscow 119334, Russia, Email: igplashchina@sky.chph.ras.ru

G. Donadze
Department of Condensed Matter Physics, E. Andronikashvili Institute of Physics, Tbilisi 0177, Republic of Georgia

R. M. Garipov
Kazan National Research Technological University, Kazan, Tatarstan, Russia

V. K. Gerasimov
A. N. Frumkin Institute of Physical Chemistry and Electrochemistry of the Russian Academy of Sciences (IPCE RAS), Moscow, 119991, Russia

M. A. Goldshtrakh
Moscow M.V. Lomonosov State University of Fine Chemical Technology, 119571 Moscow, Email: aolkhov72@yandex.ru

K. Z. Gumargalieva
N. N. Semenov Institute of Chemical Physics, RAS, Moscow, Russia

A. K. Haghi
Associate member of the Canadian Research and Development Center of Sciences and Cultures, Montreal, Quebec, Canada

A. N. Inozemtsev
M.V. Lomonosov MSU, Biological Faculty, Leninskie Gory, Moscow 119991, Russia, Email: olgakarp@newmail.ru

A. L. Iordanskii
Semenov Institute of Chemical Physics, Russian Academy of Sciences, Moscow, Russia

V. F. Kablov
Volzhsky Polytechnical Institute, (Branch of Federal State Budgetary Educational Institution of Higher Professional Education) Volgograd State Technical University, Volzhsky 404121, Russia, Email: kablov@volpi.ru; vtp@volpi.ru; d.provotorova@gmail.com; www.volpi.ru

S. G. Karpova
Emanuel Institute of Biochemical Physics, Russian Academy of Sciences, Moscow, Russia

O. V. Karpukhina
N. N. Semenov Institute of Chemical Physics, RAS, Moscow, Russia

N. A. Keibal
Volzhsky Polytechnical Institute, (Branch of Federal State Budgetary Educational Institution of Higher Professional Education) Volgograd State Technical University, Volzhsky 404121, Russia, Email: vtp@volpi.ru; d.provotorova@gmail.com; www.volpi.ru

S. V. Kolesov
The Institute of Organic Chemistry of the Ufa Scientific Centre the Russian Academy of Science, October Prospect 71, 450054 Ufa, Russia

T. V. Krekaleva
Volzhsky Polytechnical Institute, (Branch of Federal State Budgetary Educational Institution of Higher Professional Education) Volgograd State Technical University, Volzhsky 404121, Russian Federation, E-mail: vtp@volpi.ru; www.volpi.ru

E. T. Kruts'ko
Doctor of Technical Sciences, Professor (BSTU)

E. I. Kulish
Bashkir State University, Russia, Republic of Bashkortostan, Ufa, 450074, Email: alenakulish@rambler.ru

S. M. Lomakin
Emanuel Institute of Biochemical Physics, Russian Academy of Sciences, Moscow, Russia

T. Machaidze
Departmet of Condensed Matter Physics, E. Andronikashvili Institute of Physics, Tbilisi 0177, Republic of Georgia

M. Mehdipour
Textile Engineering Department, Guilan University, Rasht, Iran

F. V. Morev
Postgraduate Belarusian State Technological University, Minsk, Republic of Belarus, Email: prok_nr@mail.by

T. M. Natriashvili
Laboratory for Polymer materials, Institute of Machine Mechanics, Tbilisi 0186, Republic Georgia, Email: jimaneli@yahoo.com

A. A. Olkhov
Moscow M.V. Lomonosov State University of Fine Chemical Technology, 119571 Moscow, Email: aolkhov72@yandex.ru

I. G. Plashchina
Institute of Biochemical Physics N. M. Emanuel Academy of Sciences, Moscow 119334, Russia, Email: igplashchina@sky.chph.ras.ru

A. V. Polyakov
Institute of Biochemical Physics N. M. Emanuel Academy of Sciences, Moscow 119334, Russia, Email: igplashchina@sky.chph.ras.ru

A. A. Popov
Emanuel Institute of Biochemical Physics, Russian Academy of Sciences, Moscow, Russia

N. R. Prokopchuk
Corresponding Member of National Academy of Sciences of Belarus, Doctor of Chemical Sciences, Professor, Head of Department (BSTU)

D. A. Provotorova
Volzhsky Polytechnical Institute, branch of Federal State Budgetary Educational Institution of Higher Professional Education Volgograd State Technical University, Volzhsky 404121, Russia, Email: d.provotorova@gmail.com ; www.volpi.ru

S. V. Rudakov
State University of Moldova, MD-2009 Kishinev, Moldova.

A. S. Rudakova
State University of Moldova, MD-2009 Kishinev, Moldova.

S. N. Rusanova
Kazan National Research Technological University

N. G. Shilkina
Emanuel Institute of Biochemical Physics, Russian Academy of Sciences, Moscow, Russia

A. S. Shurshina
The Institute of Organic Chemistry of the Ufa Scientific Centre the Russian Academy of Science, October Prospect 71, Ufa 450054, Russia

A. D. Shutov
State University of Moldova, MD-2009 Kishinev, Moldova.

S. Yu. Sofina
Kazan National Research Technological University

A. G. Stepanova
Volzhsky Polytechnical Institute (branch) Volgograd State Technical University, Volzhsky 404121, Russian Federation, Email: vtp@volpi.ru ; www.volpi.ru

O. V. Stoyanov
Kazan National Research Technological University

N. E. Temnikova
Kazan National Research Technological University

R. Tukhvatullin
Bashkir State University, Ufa 450076, Republic of Bashkortostan, Russia

D. Vasiliev
Bashkir State University, Ufa 450076, Republic of Bashkortostan, Russia

A. A. Volodkin
Federal state budgetary establishment of a science of Institute of biochemical physics of N. M. Emanuelja of the Russian Academy of Sciences, Email: chembio@sky.chph.ras.ru

G. E. Zaikov
N. M. Emanuel Institute of Biochemical physics Russian Academy of Sciences, Moscow 119991, Russia, Email: Chembio@sky.chph.ras.ru;
Laboratory of Polymer Aging and Stabilization, Institute of Chemical Biophysics of Russian Academy of Sciences, Moscow, Russia

A. A. Zhivaev
Volzhsky Polytechnical Institute (branch) Volgograd State Technical University, Volzhsky 404121, Russian Federation, Email: vtp@volpi.ru; www.volpi.ru

I. L. Zhuravleva
Institute of Biochemical Physics N. M. Emanuel Academy of Sciences, Moscow 119334, Russia, Email: igplashchina@sky.chph.ras.ru

LIST OF ABBREVIATIONS

AC	activated carbon
AFM	atomic force microscopy
AM	antibiotic amikacin
BSA	bovine serum albumin
CME	clathrin-mediated endocytosis
CNR	chlorinated natural rubber
CNTs	carbon nanotubes
CS	cellulose
DSC	differential scanning calorimeter
ER	epoxy resin
EVA	ethylene with vinylacetate
Fcc	face-centered cubic
FG	fiber glass
FRC	fiber-reinforced concrete
GCMD	grand canonical molecular dynamics
HTS	high-temperature shearing
ITZ	interfacial transition zone
IUPAC	international union of pure and applied chemistry
LDPE	low-density polyethylene
MC	Monte Carlo
MD	molecular dynamics
MF	microfiltration
MNPs	magnetic nanoparticles
MS	medicinal substance
MSD	mean-square displacement
MWCO	molecular weight cut-off
MWNTs	multi-walled carbon nanotube
NC	nanocarbon
NEMD	non-equilibrium MD
NF	nanofiltration
PCTs	physical and chemical transformations
PER	phenol formaldehyde resin
PET	polyethylene
PHB	polyhydroxybutyrate
PIR	piracetam

PMS	polymethyl-silsesquioxane
PP	polypropylene
RESPA	reference system propagator algorithm
RO	reverse osmosis
RSM	response surface methodology
SLN	solid lipid nanoparticles
SWNTs	single-walled carbon nanotubes
TEM	transmission electron microscopy
TMs	technical materials
UF	ultrafiltration
VACF	velocity autocorrelation function
VCF	velocity correction factor

LIST OF SYMBOLS

X_e	equilibrium moisture content of dehydrated material
$MR_{exp,i}$	measured value at point i
n_{param}	number of parameters in the particular model
$MR_{Pred,i}$	predicted value at that point
$G_s(t)$	concentration of the desorbed substance at time t
k	constant connected with parameters of interaction polymer–diffuse substance
K	drying rate
L	length of the cylindrical sample
L	thickness of the film sample
m_∞	relative amount of water in equilibrium swelling film sample
MR	moisture ratio
n	indicator characterizing the mechanism of transfer of substance
R	correlation coefficient
R	radius of the cylindrical sample
t	time of drying
X	material moisture content

PREFACE

Through advanced characterization and new fabrication techniques, the physics, chemistry, and structure of functional materials have become a central focus of investigation in materials science, chemistry, physics, and engineering. This book presents a detailed overview of recent research developments on functional materials, including synthesis, characterization, and applications. A series of chapters provides state-of-the-art information on structures and performance of polymer composites.

This volume contains a series of topical articles by prominent leaders in this field. The research presented here discusses design principles, candidate materials and systems, and current advances and will serve as a useful source of insights into this field. This book provides a strong understanding of the primary types of materials and composites, as well as the relationships that exist between the structural elements of materials and their properties. The relationships among processing, structure, properties, and performance components are explored throughout the chapters.

CHAPTER 1

AN OVERVIEW ON NANOSYSTEMS: PROMISES AND CHALLENGES FOR THE FUTURE

A. K. HAGHI and G. E. ZAIKOV

CONTENTS

1.1   INTRODUCTION

Understanding the Nano world makes up one of the frontiers of modern science. One reason is that technology based on nanostructures promises to be largely important economically. Nanotechnology literally means any technology one nanoscale that has applications in the real world. It includes the production and application of physical, chemical, and biological system sat scales ranging from individual atoms or molecules to submicron dimensions, as well as the integration of the resulting nanostructures into larger systems. Nanotechnology probably has a profound impact on our economy and society in the early twenty-first century, comparable to that of semiconductor technology, information technology, or cellular and molecular biology. Science and technology research in nanotechnology promises breakthroughs in areas such as materials and manufacturing, nanoelectronics, medicine and healthcare, energy, biotechnology, information technology, and national security. It is widely felt that nanotechnology will be the next Industrial revolution.

Nanotechnology is considered one of the most promising technologies of the twenty-first century.

Nanotechnology expands itself as one of the major areas of science [1–25]. An important and interesting aspect of nanomedicine is the use of nanoparticle for target specific drug-delivery systems, and it allows a new innovative therapeutic approaches. Due to their small size, these drug-delivery systems are promising tools in therapeutic approaches, such as selective or targeted drug-delivery toward a specific tissue or organ. They also enhance drug transport across biological barriers and intracellular drug-delivery. Nanotherapeutic agents work successfully against traditional drugs, which exhibit very less bioavailability and are often associated with gastrointestinal side effects. Nanotherapeutics improves the delivery of drugs that cannot regularly be taken orally, and it also improves the safety and efficacy of low-molecular-weight drugs. It also improves the stability and absorption of proteins that normally cannot be taken orally.

1.2   PROPERTIES OF NANOTHERAPEUTICS

A good therapeutic agent or nanoparticle should have the ability to (i) cross one or various biological membranes (e.g., mucosa, epithelium, and endothelium) before, (ii) diffusing through the plasma membrane to, and (iii) finally gain access to the appropriate organelle where the biological target is located.

Different methods of Drug delivery:

(1) Oral drug-delivery
(2) Injection-based drug-delivery
(3) Transdermal drug-delivery
(4) Bone marrow infusion
(5) Control release systems
(6) Targeted drug-delivery:
    (a) Therapeutically monoclonal
    (b) Antibodies
    (c) Liposomes
    (d) Micro-particles
    (e) Modified blood cells
    (f) Nanoparticles
(7) Implant drug-delivery system

1.3   METAL-BASED NANOPARTICLES

Metallic nanoparticles are either spherical metal or semiconductor particles with nanometer-sized diameters. Similar to nanoparticles, "nanocrystal" or "nanocrystallite" term is used for nanoparticles made up of semiconductors. Generally, semiconductor materials from the groups II-VI (CdS, CdSe, CdTe), III-V (InP, InAs), and IV-VI (PbS, PbSe, PbTe) are of particular interest. Metal-based nanoparticles are used for nanoscale transistors, biological sensors, and next-generation photovoltaics [26–49].

1.4   LIPID-BASED NANOPARTICLES

Lipid-based nanoparticles are widely used for drug targeting and drug delivery. Solid lipid nanoparticles (SLN) introduced in 1991 represent an alternative carrier system for drugs [4]. Nanoparticles made from solid lipids attract major attention as novel colloidal drug carrier for intravenous applications as they have been proposed as an alternative particulate carrier system. SLN are submicron colloidal carriers ranging from 50 to 1,000 nm, which are composed of physiological lipid [23–52].

Lipid-based systems are mainly used because of the following reasons:

(1) They enhance oral bioavailability and reduce plasma profile variability.

(2) They have better characterization of lipoid excipients.

(3) They have an improved ability to address the key issues of technology transfer and manufacture scale-up.

1.5   POLYMER-BASED NANOPARTICLES

Polymeric nanoparticles are nanoparticles that are prepared from polymers. Polymeric nanoparticles form (i) the micronization of a material into nanoparticles and (ii) the stabilization of the resultant nanoparticles. As for the micronization, one can start with either small monomers or a bulk polymer. The drug is dissolved, entrapped, encapsulated, or attached to a nanoparticle and one can obtain different nanoparticles, nanospheres, or nanocapsules according to the methods of preparation. Gums, gelatin sodium alginate albumin are used for polymer-based drug-delivery. Polymeric nanoparticles are prepared by cellulosics, poly(2-hydroxy ethyl methacrylate), poly(N-vinyl pyrrolidone), poly(vinyl alcohol), poly(methyl methacrylate), poly(acrylic acid), polyacrylamide, and poly(ethylene-covinyl acetate)-like polymeric materials. Polymer used in drug delivery must have the following qualities such as it should be chemically inert, nontoxic, and free of leachable impurities [39–60].

1.6   BIOLOGICAL NANOPARTICLES

For proper interaction with molecular targets, a biological or molecular coating or layer is required, which can act as a bioinorganic interface, and it should be attached to the nanoparticle for proper interaction. These biological coatings may be antibodies and biopolymers such as collagen. A layer of small molecules that make the nanoparticles biocompatible are also used as bioinorganic interface. For detection and analysis, these nanoparticles should be optically active and they should either fluoresce or change their color in different intensities of light.

1.7 NANOCARRIERS IN DRUG DELIVERY

Nanoparticles and liposomes get absorbed by blood streams for its mode of action. For particular function these nanoparticles must have specific structure and composition. These particles rapidly get cleared from blood stream by macrophages of reticuloendothelial system [13–28].

Nanoparticles can be delivered to targets by

PHAGOCYTOSIS PATHWAY

Phagocytosis occurs by professional phagocytes and by nonprofessional phagocytes. A professional phagocyte includes macrophages, monocytes, neutrophils, and dendritic cells, while nonprofessional phagocytes are fibroblast, epithelial, and endothelial cells.

Phagocytosis occurs due to

(i) Recognition of the opsonized particles in the blood stream

(ii) Adhesion of particles to macrophages

(iii) Ingestion of the particle

Opsonization of the nanoparticlesis is a major process and takes place before phagocytosis. In this process, nanoparticles get tagged by a protein known as psonins. Due to this tagging, nanoparticles make a visible complex for macrophages. This whole process takes place in blood streams.

These activated particles then get attached to macrophages and this interaction is similar to receptor—ligand interaction [61–80].

As actin is depolymerized from the phagosome, the newly denuded vacuole membrane becomes accessible to early endosomes. Through a series of fusion and fission events, the vacuolar membrane and its contents will mature, fusing with late endosomes and ultimately lysosomes to form a phagolysosome (Figure 1.1). The rate of such events depends on the surface properties of the ingested particle, typically from half to several hours. The phagolysosomes become acidified due to the vacuolar proton pump ATPase located in the membrane and acquire many enzymes, including esterases and cathepsins [81–95].

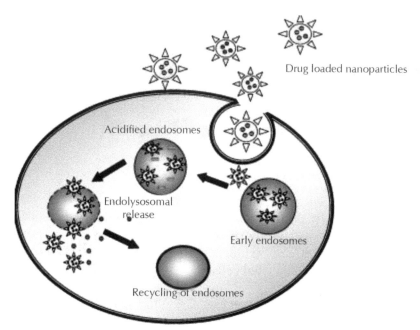

FIGURE 1.1   Nanoparticles enter the cell through receptor-mediated endocytosis and get localized in the endolysosomal compartment.

NONPHAGOCYTIC PATHWAY

This is also known as pinocytosis or cell-drinking method. This is basically imbibing of fluids and solutes. Very small nanoparticles enter in well and get absorbed by pinocytosis as phagocytosis are restricted to the size of particle. The pinocytosis process occurs in specialized cells. It may be clathrin-mediated endocytosis, caveolae-mediated endocytosis, macropinocytosis, and caveolae-independent endocytosis.

1.8   CLATHRIN-MEDIATED ENDOCYTOSIS

Endocytosis via clathrin-coated pits, or clathrin-mediated endocytosis (CME), occurs constitutively in all mammalian cells and fulfills crucial physiological roles, including nutrient uptake and intracellular communication. For most cell types, CME serves as the main mechanism of internalization for macromolecules and plasma membrane constituents. CME via specific receptor—ligand interaction is the best described mechanism, to the extent that it was previously referred to as "receptor-mediated endocytosis." However, it is now clear that alternative nonspecific endocytosis via clathrin-coated pits also exists (as well as receptor-mediated but clathrin-independent endocytosis). Notably, the CME, either receptor-dependent or independent, causes the endocytosed material to end up in degradative lysosomes. This has an important impact in the drug-delivery field, as the drug-loaded nanocarriers may be tailored in order to become metabolized into the lysosomes, thus releasing their drug content intracellularly as a consequence of lysosomal biodegradation [96–103].

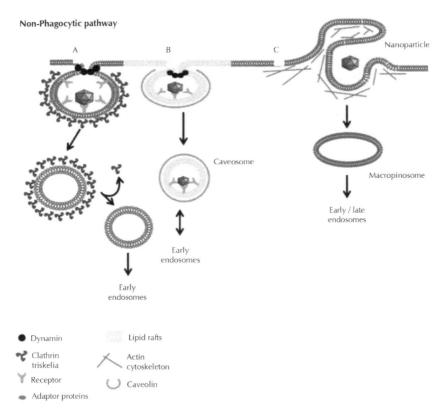

FIGURE 1.2 Diagrammamticpresentation of (A) clathrin-mediated endocytosis, (B) caveolae-mediated endocytosis, and (C) macropinocytosis.

1.9   CAVEOLAE-MEDIATED ENDOCYTOSIS

Although CME is the predominant endocytosis mechanism in most cells, alternative pathways have been more recently identified, caveolae-mediated endocytosis (CvME) being the major pathway. Caveolae are characteristic flask-shaped membrane invaginations, with a size generally reported in the lower end of the 50–100 nm range, typically 50–80 nm. They are lined by caveolin, a dimeric protein, and enriched with cholesterol and sphingolipids (Figure 1.2). Caveolae are particularly abundant in endothelial cells, where they can constitute 10–20 per cent of the cell surface, but also smooth muscle cells and fibroblasts. CvMEs are involved in endocytosis and trancytosis of various proteins; they also constitute a port of entry for viruses (typically the SV40 virus) and receive increasing attention for drug-delivery applications using nanocarriers.

Unlike CME, CvME is a highly regulated process involving complex signaling, which may be driven by the cargo itself. After binding to the cell surface, particles move along the plasma membrane to caveolae invaginations, where they may be maintained through receptor—ligand interactions. Fission of the caveolae from the membrane, mediated by the GTPasedynamin, then generates the cytosolic caveolar vesicle, which does not contain any enzymatic cocktail. Even this pathway is employed by many pathogens to escape degradation by lysosomal enzymes. The use of nanocarriers exploiting CvME may therefore be advantageous to bypass the lysosomal degradation pathway when the carried drug (e.g., peptides, proteins, nucleic acids, etc.) is highly sensitive to enzymes. On the whole, the uptake kinetics of CvME is known to occur at a much slower rate than that of CME. Ligands known to be internalized by CvME include folic acid, albumin, and cholesterol.

1.10   MACROPINOCYTOSIS

Macropinocytosis is another type of clathrin-independent endocytosis pathway, occurring in many cells, including macrophages. It occurs through the formation of actin-driven membrane protrusions, similar to phagocytosis. However, in this case, the protrusions do not zipper up along the ligand-coated particle; instead, they collapse onto and fuse with the plasma membrane. This generates large endocytic vesicles, called macropinosomes, which sample the extracellular milieu and have a size generally larger than 1 lm (and sometimes as large as 5 lm). The intracellular fate of macropinosomes vary depending on the cell type, but in most cases, they acidify and shrink. They may eventually fuse with lysosomal compartments or recycle their content to the surface (Figure 1.2). Macropinosomes have not been reported to contain any specific coating, nor do they concentrate receptors. This endocytic pathway does not seem to display any selectivity, but is involved, among others, in the uptake of drug nanocarriers [80–103].

1.11   MAGNETIC NANOPARTICLES

Magnetic nanoparticles (MNPs) have many applications in different areas of biology and medicine. MNPs are used for hyperthermia, magnetic resonance imaging, immunoassay, purification of biologic fluids, cell and molecular separation, and tissue engineering. The design of magnetically targeted nanosystems (MNSs) for a smart delivery of drugs to target cell-sis a promising direction of nanobiotechnology. They traditionally consist of one or more magnetic cores and biological or synthetic molecules that serve as a basis for polyfunctional coatings on MNPs surface. The coatings of MNSs should meet several important requirements. They should be biocompatible, protect magnetic cores from influence of biological liquids, prevent MNSs' agglomeration in dispersion, provide MNSs localization in biological targets and homogeneity of MNSs' sizes. The coatings must be fixed on MNPs' surface and contain therapeutic products (drugs or genes) and biovectors for recognition by biological systems. The model that is often used when MNSs are developed is presented in (Figure 1.3).

Proteins are promising materials for creation of coatings on MNPs for biology and medicine. When proteins are used as components of coatings, it is of utmost importance that they keep their functional activity. Protein binding on MNPs' surface is a difficult scientific task. Traditionally bifunctional linkers (glutaraldehyde, carbodiimide) are used for protein cross-linking on the surface of MNPs and modification of coatings by therapeutic products and biovectors. Researchers modified MNPs' surface with aminosilanes and performed protein molecule attachment using glutaraldehyde. In the issue bovine serum albumin (BSA) was adsorbed on MNPs' surface in the presence of carbodiimide. These works revealed several disadvantages of this method of protein fixing, which make it un-promising. Some of these are cluster formations as a result of linking of protein molecules adsorbed on different MNPs, desorption of proteins from MNSs' surface as a result of incomplete linking, and uncontrollable linking of proteins in solution (Figure 1.4). The creation of stable protein coatings with retention of native properties of molecules is still an important biomedical issue.

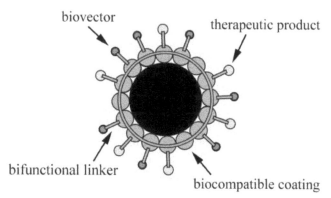

FIGURE 1.3 The classical scheme of magnetically targeted nanosystem for a smart delivery of therapeutic products.

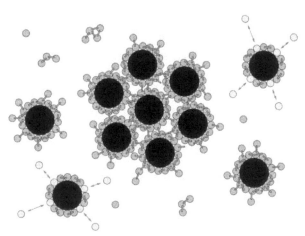

FIGURE 1.4 Nonselective linking of proteins on MNPs surface by bifunctional linkers leading to clusters formation and desorption of proteins from nanoparticles surface.

It is known that proteins can be chemically modified in the presence of free radicals with formation of cross-links. The goals of the work were to create stable protein coating on the surface of individual MNPs using a fundamentally novel approach based on the ability of proteins to form an interchange of covalent bonds under the action of free radicals and estimate activity of proteins in the coating.

1.12 EXPERIMENTAL

1.12.1 MAGNETIC SORBENT SYNTHESIS

Nanoparticles of magnetite Fe3O4 were synthesized by coprecipitation of ferrous and ferric salts in water solution at 4°C and *in the alkaline medium*:

$$Fe^{2+} + 2Fe^{3+} + 8OH^- \rightarrow Fe_3O_4\downarrow + 4H_2O$$

Around 1.4 g of $FeSO_4 \cdot 7H_2O$ and 2.4 g of $FeCl_3 \cdot 6H_2O$ were dissolved in 50 ml of distilled water so molar ratio of Fe^{2+}/Fe^{3+} willbe equal to 1:2. After filtration of the solution, 10 ml of 25 mass per cent NH_4OH was added on a magnetic stirrer; 2.4 g of PEG 2 kDa) was added previously in order to reduce the growth of nanoparticles during the reaction. After the precipitate was formed, the solution (with 150 ml of water) was placed on a magnet. Magnetic particles precipitated on it and supernatant liquid was deleted. The procedure of particle washing was repeated for 15 times until neutral pH was obtained. MNPs were stabilized by double electric layer with the use of disperser. To create the double electric layer, 30ml of 0.1 M phosphate-citric buffer solution (0.05M NaCl) with pH value of 4 was introduced. MNPs' concentration in hydrosol was equal to 37 mg/ml.

1.12.2 PROTEIN COATING FORMATION

BSA and thrombin with activity of 92 units per cent 1 mg were used for protein coating formation. Several types of reaction mixtures were used: "A1-MNP-0," "A2-MNP-0," "A1-MNP-1," "A2-MNP-1," "A2-MNP-1-acid," "T1-MNP-0," "T1-MNP-0," and "T1-0-0." All of them contained
(1) 2.80 ml of protein solution ("A1" or "A2" means that there is BSA solution with concentration of 1 mg/ml or 2 mg/ml in 0.05 M

phosphate buffer with pH 6.5 (0.15M NaCl) in the reaction mix-ture; "T1" means that there is thrombin solution with concentration of 1 mg/ml in 0.15M NaCl with pH 7.3);

(2) 0.35 ml of 0.1 M phosphate-citric buffer solution (0.05 M NaCl) or MNPs hydrosol ("MNP" in the name of reaction mixture means that it contains MNPs);

(3) 0.05 ml of distilled water or 3 mass per cent H_2O_2 solution ("0" or "1" in the reaction mixture names correspondingly).

Hydrogen peroxide interacts with ferrous ion on MNPs' surface with formation of hydroxylradicals by Fenton reaction:

$$Fe^{2+} + H_2O_2 \rightarrow Fe^{3+} + OH^{\cdot} + OH^-$$

"A2-MNP-1-acid" is a reaction mixture, containing 10 μl of ascorbic acid with a concentration of 152 mg/ml. Ascorbic acid is known to form free radicals in reaction with H_2O_2 and generate free radicals in solution but not only on MNPs' surface.

The sizes of MNPs, proteins, and MNPs in adsorption layer were ana-lyzed using dynamic light scattering (Zetasizer Nano S "Malvern," Eng-land) with detection angle of 173° at temperature 25°C.

1.12.3   STUDY OF PROTEIN ADSORPTION ON MNPS

The study of protein adsorption on MNPs was performed using ESR-spec-troscopy of spinlabels. The stable nitroxide radical used as spin label is presented in **Figure (1.5)**. Spin label technique allows studying adsorption of macromolecules on nanosized magnetic particles in dispersion without complicated separation processes of solution components. The principle of quantitative evaluation of adsorption is as follows: Influence of local fields of MNPs on spectra of radicals in solution depends on the distance between MNPs and radicals. If this distance is lower than 40 nm for mag-netite nanoparticles with an average size of 17 nm, ESR spectra lines of the radicals broaden strongly and their intensity decreases to zero. The decrease in the spectrum intensity is proportional to the part of radicals that are located inside the layer of 40 nm in thickness around MNP. The same occurs with spin labels covalently bound to protein macromolecules. The intensity of spin labels spectra decreases as a result of adsorption of macromolecules on MNPs (Figure 1.6). We have shown that spin labels technique can be used for the study of adsorption value, adsorption kinetics, calculation of average number of molecules in adsorption layer and adsorption layer thickness, and concurrent adsorption of macromolecules.

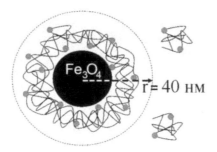

FIGURE 1.5   The stable nitroxide radical used for labeling of macromolecules containing amino groups (1) and spin label attached to protein macromolecule (2).

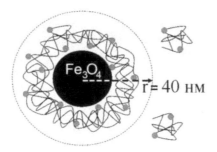

FIGURE 1.6   Magnetic nanoparticle sand spin-labeled macromolecules in solution.

The reaction between the radical and protein macromolecules was conducted at room temperature. A total of 25 µl of radical solution in 96 per cent ethanol with concentration of 2.57 mg/ml was added to 1 ml of protein solution. The solution was incubated for 6 h and dialyzed. The portion of adsorbed protein was calculated from intensity of the low-field line of nitroxide radical triplet I_{+1}.

The method of ferromagnetic resonance was also used to study adsorption layer formation.

The spectra of the radicals and MNPs were recorded at room temperature using spectrometer at a microwave power of 5 mW, modulation frequency 100 kHz, and amplitude 1 G. The first derivative of the resonance absorption curve was detected. The samples were placed into the cavity of the spectrometer in a quartz flat cell. Magnesium oxide powder containing Mn^{2+} ions was used as an external standard in ESR experiments. Average amount of spin labels on protein macromolecules reached 1 per cent 4–5 albumin macromolecules and 1 per cent 2–3 thrombin macromolecules. Rotational correlation times of labels were evaluated as well as a fraction of labels with slow motion.

1.12.4   COATING STABILITY ANALYSIS AND ANALYSIS OF SELECTIVITY OF FREE RADICAL PROCESS

In our previous works, it was shown that fibrinogen (FG) adsorbed on MNPs' surface forms thick coating and micronsized structures. Also, FG demonstrates an ability to replace BSA previously adsorbed on MNPs surface. This was proved by complex study of systems containing MNPs, spin-labeled BSA and FG with spin labels technique and ferromagnetic resonance. The property of FG to replace BSA from MNPs surface was used in this work for estimating BSA coating stability. A total of 0.25 ml of FG solution with concentration of 4 mg/ml in 0.05 M phosphate buffer with pH 6.5 was added to 1 ml of the samples "A1-MNP-0," "A2-MNP-0," "A1-MNP-1,"and "A2-MNP-1." The cluster formation was observed by dynamic light scattering.

The samples "A2-MNP-0," "A2-MNP-1," "T1-MNP-0,"and "T1-MNP-1" were centrifuged at 1,20,000 g for 1 h. On these conditions, the precipitation of MNPs occurs, but macromolecules physically adsorbed on MNPs remain as supernatant liquid. The precipitates containing MNPs and protein fixed on MNPs' surface were dissolved in buffer solution with subsequent evaluation of the amount of protein by Bradford colorimetric method. Also, spectrophotometer was used.

Free radical modification of proteins in supernatant liquids of "A2-MNP-0" "A2-MNP-1," and an additional sample of "A2-MNP-1-acid" were analyzed by IR-spectroscopy using FTIR-spectrometer with DTGS-detector with 2 cm^{-1} resolution. Comparison of "A2-MNP-0," "A2-MNP-1," and "A2-MNP-1-acid" were used to reveal the selectivity of free radical process in "A2-MNP-1."

1.12.5   ENZYME ACTIVITY ESTIMATION

Estimation of enzyme activity of protein fixed on MNPs' surface was performed on the example of thrombin. This protein is a key enzyme of blood-clotting system that catalyzes the process of conversion of fibrinogen to fibrin. Thrombin may lose its activity as a result of free radical modification, and the rate of the enzyme reaction may decrease. Thus, estimation of enzyme activity of thrombin cross-linked on MNPs' surface during free radical modification was performed by comparison of the rates of conversion of fibrinogen to fibrin under the influence of thrombin contained in reaction mixtures. A total of 0.15 ml of the samples "T1-MNP-0," "T1-MNP-1," and "T1-0-0" was added to 1.4 ml of FG solution at a concentration of 4 mg/ml. Kinetics of fibrin formation was studied by Rayleigh light scattering on spectrometer with multi bit 64-channel correlator.

1.13   RESULTS AND DISCUSSION

ESR spectra of spin labels covalently bound to BSA and thrombin macromolecules (Figure 1.7) allow obtaining information on their microenvironment. The spectrum of spin labels bound to BSA is a superposition of narrow and wide lines characterized by rotational correlation times of 10^{-9} and $2 \cdot 10^{-8}$ s. This is an evidence of existence of two main regions of spin labels localization on BSA macromolecules. The portion of labels with slow motion is about 70 per cent. Thus, a considerable part of labels is situated in internal areas of macromolecules with high micro viscosity. The labels covalently bound to thrombin macromolecules are characterized by one rotational correlation time of 0.26 ns. These labels are situated in areas with equal micro viscosity.

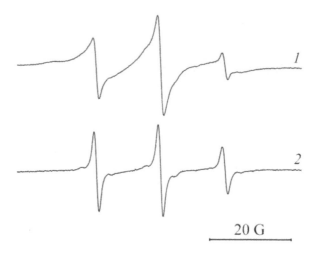

FIGURE 1.7   ESR spectra of spin labels on BSA (1) thrombin and (2) at 25°C.

The signal intensity of spin-labeled macromolecules decreased after introduction of MNPs into the solution *that testifies to the* protein adsorption on MNPs (Figure 1.8). Spectra of the samples "A1-MNP-0" and "T1-MNP-0" consist of nitroxide radical triplet, the third line of sextet of Mn^{2+} (the external standard) and ferromagnetic resonance spectrum of MNPs. Rotational correlation time of spin labels does not change after MNPs addition. The dependences of spectra lines intensity for spin-labeled BSA and thrombin in the presence of MNPs on incubation time are summarized in Table (1.1). Signal intensity of spin-labeled BSA changes insignificantly. These changes correspond to adsorption of approximately 12 per cent of BSA after the sample incubation for 100 min. The study of adsorption kinetics reveals that establishing that adsorption equilibrium in "T1-MNP-0" should take place when the incubation time equals 80 min and ~41 per cent of thrombin is adsorbed. The value of adsorption A may be estimated using the data on the portion of macromolecules adsorbed and specific surface area calculated from MNPs density (5,200 mg/m^3), concentration, and size. Hence, BSA adsorption equals to 0.35 mg/m^2 after 100 min of incubation. The dependence of thrombin adsorption value on incubation time is shown in Figure (1.9). Thrombin adsorption equals to 1.20 mg/m^2 after 80 min of incubation.

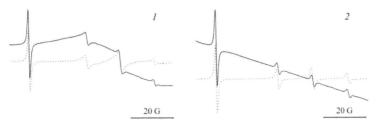

FIGURE 1.8   ESR spectra of spin labels on BSA (1) thrombin and (2) macromolecules before (dotted line) and 75 min after (solid line) MNPs addition to protein solution at 25°C. External standard—MgO powder containing Mn^{2+}.

TABLE 1.1   The dependence of relative intensity of low-field line of triplet I_{+1} of nitroxide radical are covalently bound to BSA and thrombin macromolecules, and the portion N of the protein is adsorbed on incubation time t of the samples "A1-MNP-0" and "T1-MNP-0."

	Spin-labeled BSA		Spin-labeled thrombin	
t, min.	I_{+1}, rel. units	N, %	I_{+1}, rel. units	N, %
0	0.230 ± 0.012	0 ± 5	0.25 ± 0.01	0 ± 4
15	-	-	0.17 ± 0.01	32 ± 4
35	0.205 ± 0.012	9 ± 5	0.16 ± 0.01	36 ± 4
75	0.207 ± 0.012	10 ± 5	0.15 ± 0.01	40 ± 4
95	-	-	0.15 ± 0.01	40 ± 4
120	0.200 ± 0.012	13 ± 5	0.14 ± 0.01	44 ± 4

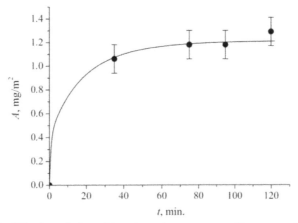

FIGURE 1.9   Kinetics of thrombin adsorption on magnetite nanoparticles at 25°C. Concentration of thrombin in the sample is 0.9 mg/ml, MNPs—4.0 mg/ml.

The FMR spectra of the samples "A1-MNP-0," "T1-MNP-0," and MNPs are characterized by different positions in the magnetic field (Figure 1.10). The center of the spectrum of MNPs is 3254 G, whereas the centers of "A1-MNP-0" and "T1-MNP-0" spectra are 3253 G and 3449 G, respectively. Resonance conditions for MNPs in magnetic field of spectrometer include a parameter of the shift of FMR spectrum $|M_1| = \frac{3}{2}|H_1|$, where H_1 is a local field created by MNPs in linear aggregates, which form in spectrometer field. $H_1 = 2\sum_{1}^{\infty} \frac{2\mu}{(nD)^3}$, where D is a distance between MNPs in linear aggregates, μ is MNPs' magnetic moment, and n is number of MNPs in aggregate. Coating formation and the thickness of adsorption layer decrease dipole interactions and particles ability to aggregate. As a result, the center of FMR spectrum moves to higher fields. This phenomenon of FMR spectrum center shift is observed in the system "A1-MNP-0" after FG addition. The spectrum of MNPs with thick coating becomes similar to FMR spectra of isolated MNPs. Thus, the similar center positions of FMR spectra of MNPs without coating and MNPs in BSA coating point to a very thin coating and low adsorption of protein in this case. In contrast, according to FMR center positions, the thrombin coating on MNPs is thicker than albumin coating. *This result is consistent with the data* obtained by ESR spectroscopy.

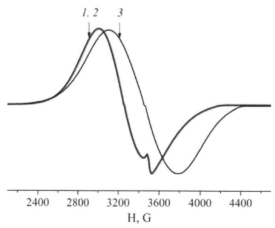

FIGURE 1.10 (1) FMR spectra of MNPs, (2) MNPs in the mixture with BSA (the sample "A1-MNP-0") after incubation time of 120 min, and (3) MNPs in the mixture with thrombin (the sample "T1-MNP-0") after incubation time of 120 min.

FG's ability to replace BSA in adsorption layer on MNPs surface is demonstrated in Figure (1.11). Initially, there is bimodal volume distribution of particles over sizes in the sample "A2-MNP-0" that can be explained by existence of free (unadsorbed) BSA and MNPs in BSA coating. After FG addition the distribution changes. Micronsized clusters form in the sample that proves FG adsorption on MNPs [18]. In the case of "A2-MNP-1," volume distribution is also bimodal. The peak of MNPs in BSA coating is characterized by particle size of maximal contribution to the distribution of ~23 nm. This size is identical to MNPs in BSA coating in the sample "A2-MNP-0." This proves that H_2O_2 addition does not lead to uncontrollable linking of protein macromolecules in solution or cluster formation. As MNPs' size is 17 nm, the thickness of adsorption layer on MNPs is approximately 3 nm.

After FG addition to "A2-MNP-1," micronsized clusters do not form. Thus, adsorption BSA layer formed in the presence of H_2O_2 keeps stability. This stability can be explained by the formation of covalent bonds between protein macromolecules [13] in adsorption layer as a result of free radical generation on MNPs' surface. Stability of BSA coating on MNPs was demonstrated for the samples "A1-MNP-1" and "A2-MNP-1" incubated for more than 100 min before addition of FG. Clusters are shown to appear if the incubation time is insufficient.

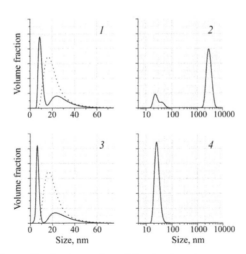

FIGURE 1.11   Volume distributions of particles in sizes in systems without (1, 2) and with (3, 4) H_2O_2 ("A2-MNP-0", "A2-MNP-1") incubated for 2 h before (1, 3) and 20 min after (2, 4) FG addition. Dotted line is the volume distribution of nanoparticles in sizes in dispersion.

The precipitates obtained by ultracentrifugation of "A2-MNP-0," "A2-MNP-1," "T1-MNP-0," and "T1-MNP-1" were dissolved in buffer solution. The amount of protein in precipitates was evaluated using colorimetric method (**Table 1.2**). The results revealed that precipitates of systems with H_2O_2 contained more protein than the same systems without H_2O_2. Therefore, in the samples containing H_2O_2 the significant part of protein molecules does not leave MNPs' surface when centrifuged, while in the samples "A2-MNP-0" and "T1-MNP-0," most part of protein molecules leave the surface. This indicates the stability of adsorption layer formed in the presence of free radical generation initiator and proves cross-link formation.

TABLE 1.2   The amount of protein in precipitates after centrifugation of the samples "A2-MNP-0", "A2-MNP-1", "T1-MNP-0" and "T1-MNP-1" of 3.2 ml in volume

Sample name	Amount of protein in precipitates, mg
"A2-MNP-0"	0.05
"A2-MNP-1"	0.45
"T1-MNP-0"	0.15
"T1-MNP-1"	1.05

Analysis of content of supernatant liquids obtained after ultracentrifugation of reaction systems containing MNPs and BSA, which differed by H_2O_2 and ascorbic acid presence ("A2-MNP-0," "A2-MNP-1," and "A2-MNP-1-acid") allows evaluating the scale of free radical processes in the presence of H_2O_2. As was mentioned above in the presence of ascorbic acid, free radicals generate not only on MNPs' surface but also in solution. Thus, both molecules on the surface and free molecules in solution can undergo free radical modification in this case. From Figure 1.12, we can see that the IR-spectrum of "A2-MNP-1-acid" differs from the spectra of "A2-MNP-0" and "A2-MNP-1," while *the spectra of* "A2-MNP-0" and "A2-MNP-1" *almost* have no differences. The IR-spectra differ in the region of 1,200–800cm^{-1}. The changes in this area are explained by free radical oxidation of amino-acid residues of methionine, tryptophane, histidine, cysteine, and phenylalanine. These residues are sulfur-containing and cyclic, which are the most sensitive to free radical oxidation. The absence of differences in "A2-MNP-0" and "A2-MNP-1" proves that cross-linking of protein molecules in the presence of H_2O_2 is selective and takes place only on MNPs' surfaces.

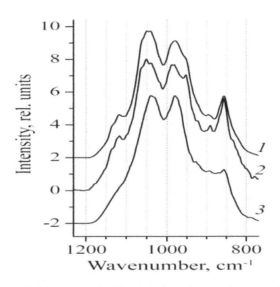

FIGURE 1.12   IR-spectra of supernatant solutions obtained after centrifugation of the samples (1) "A2-MNP-0", (2) "A2-MNP-1", and (3) "A2-MNP-1-acid".

When proteins are used as components of coating on MNPs for biology and medicine, their functional activity retaining is very important. Proteins fixed on MNPs can lose their activity as a result of adsorption on MNPs or free radical modification, which is cross-linking and oxidation, but it was shown that they do not lose it. Estimation of enzyme activity of thrombin cross-linked on MNPs' surface was performed by comparison of the rates of conversion of fibrinogen to fibrin under the influence of thrombin contained in reaction mixtures "T1-MNP-0," "T1-MNP-1," and "T1-0-0". The curves for the samples containing thrombin and MNPs, which differ by the presence of H_2O_2, had no fundamental differences that illustrate preservation of enzyme activity of thrombin during free radical cross-linking on MNPs' surface. Fibrin gel was formed during ~15min in both the cases. Rayleigh light scattering intensity was low when "T1-0-0" was used and small fibrin particles were formed in this case. The reason of this phenomenon is autolysis (self-digestion) of thrombin. Enzyme activity of thrombin, one of serine proteinases, decreases spontaneously in solution. Thus, the proteins can sustain their activity longer when adsorbed on MNPs. This way, the method of free radical cross-linking of proteins seems promising for enzyme immobilization.

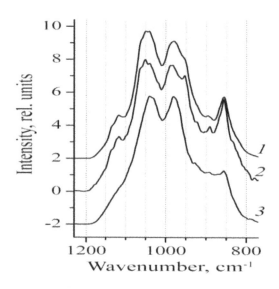

FIGURE 1.12   Kinetics curves of growth of Rayleigh light scattering intensity in the process of fibrin gel formation in the presence of (1) "T1-MNP-0", (2) "T1-MNP-1", and (3) "T1-0-0".

1.14   TRANSFER OF HEAT IN A NANOSYSTEM

Heat and mass transfer in wet nanostructure are coupled in a complicated manner. The structure of the solid matrix varies widely in shape. There is, in general, a distribution of void sizes, and the nanostructures may also be locally irregular. Energy transport in such a medium occurs by conduction in all of the phases. Mass transport occurs within voids of the medium. In an unsaturated state, these voids are partially filled with a liquid, whereas the rest of the voids contain some gas. It is a common misapprehension that nonhygroscopic fibers (i.e., those of low intrinsic for moisture vapor) will automatically produce a hydrophobic fabric. The major significance of the fine geometry of a nanostructure in contributing to resistance to water penetration can be stated in the following manner.

For instance, the requirements of a water repellent fabric are that (*a*) the nanofibers shall be spaced uniformly and as far apart as possible and (*b*) they should be held so as to prevent their ends drawing together. Meanwhile, wetting takes place more readily on surfaces of high fiber density and in a fabric where there are regions of high fiber density such as yarns, the peripheries of the yarns will be the first areas to wet out and when the peripheries are wetted, water can pass unhindered through the nanofabric.

For thermal analysis of wet nanofabrics, the liquid is water and the gas is air. Evaporation or condensation occurs at the interface between the water and air so that the air is mixed with water vapor. A flow of the mixture of air and vapor may be caused by external forces, for instance, by an imposed pressure difference. The vapor will also move relative to the gas by diffusion from regions where the partial pressure of the vapor is higher to those where it is lower.

Again, heat-transfer by conduction, convection, and radiation and moisture transfer by vapor diffusion are the most important mechanisms in very cool or warm environments from the skin.

Meanwhile, nanotextile manufacturing involves a crucial energy-intensive drying stage at the end of the process to remove moisture left from dye setting. Determining drying characteristics for nanotextiles, such as temperature levels, transition times, total drying times, and evaporation rates, is vitally important so as to optimize the drying stage. In general, drying means to make free or relatively free from a liquid. We define it more narrowly as the vaporization and removal of water from textiles.

1.14.1   HEAT

When a wet nanofabric is subjected to thermal drying, the two following processes occur simultaneously:

(a) Transfer of heat to raise the wet nanofabric temperature and to evaporate the moisture content.

(b) Transfer of mass in the form of internal moisture to the surface of the nanofabric and its subsequent evaporation.

The rate at which drying is accomplished is governed by the rate at which these two processes proceed. Heat is a form of energy that can transfer across the boundary of a system. Heat can, therefore, be defined as "the form of energy that transfers between a nanosystem and its surroundings as a result of a temperature difference." There can only be a transfer of energy across the boundary in the form of heat if there is a temperature difference between the system and its surroundings. Conversely, if the nanosystem and surroundings are at the same temperature, heat-transfer cannot take place across the boundary.

Strictly speaking, the term "heat" is a name given to a particular form of energy crossing the boundary. However, heat is more usually referred to in thermodynamics through the term "heat-transfer," which is consistent with the ability of heat to raise or lower the energy within a nanosystem.

The three modes of heat-transfer are as follows:
• Convection
• Conduction
• Radiation

All three are different. Convection relies on movement of a fluid. Conduction relies on transfer of energy between molecules within a solid or fluid. Radiation is a form of electromagnetic energy transmission and is independent of any substance between the emitter and receiver of such energy. However, all the three modes of heat-transfer rely on a temperature difference for the transfer of energy to take place.

The greater the temperature difference, the more rapid the heat will be transferred. Conversely, the lower the temperature difference, the slower will be the rate at which heat is transferred. When discussing the modes of heat-transfer, it is the rate of heat-transfer Q that defines the characteristics rather than the quantity of heat.

As it was mentioned earlier, there are three modes of heat-transfer: convection, conduction, and radiation. Although two, or even all three, modes of heat-transfer may be combined in any particular thermodynamic situation, the three are quite different and will be introduced separately.

The coupled heat and liquid moisture transport of nanoporous material has wide industrial applications in textile engineering and functional design of apparel products. Heat-transfer mechanisms in nanoporous textiles include conduction by the solid material of fibers, conduction by intervening air, radiation, and convection. Meanwhile, liquid and moisture transfer mechanisms include vapor diffusion in the void space and moisture sorption by the fiber, evaporation, and capillary effects. Water vapor moves through textiles as a result of water vapor concentration differences. Nanofibers absorb water vapor due to their internal chemical compositions and structures. The flow of liquid moisture through the textiles is caused by fiber—liquid molecular attraction at the surface of fiber materials, which is determined mainly by surface tension and effective capillary pore distribution and pathways. Evaporation and/or condensation takes place, depending on the temperature and moisture distributions. The heat-transfer process is coupled with the moisture transfer processes with phase changes such as moisture sorption and evaporation.

Mass transfer in the drying of a wet nanofabric depends on two mechanisms: movement of moisture within the fabric, which will be a function of the internal physical nature of the solid and its moisture content; and the movement of water vapor from the material surface as a result of water vapor from the material surface as a result of external conditions of temperature, air humidity and flow, area of exposed surface, and supernatant pressure.

1.14.2   CONVECTION HEAT-TRANSFER

A very common method of removing water from textiles is convective drying. Convection is a mode of heat-transfer that takes place as a result of motion within a fluid. If the fluid starts at a constant temperature and the surface is suddenly increased in temperature to above that of the fluid, there will be convective heat-transfer from the surface to the fluid as a result of the temperature difference. Under these conditions, the temperature difference causing the heat-transfer can be defined as

ΔT = (surface temperature) − (mean fluid temperature)

Using this definition of the temperature difference, the rate of heat-transfer due to convection can be evaluated using Newton's law of cooling:

$$Q = h_c A \Delta T \qquad (1.1)$$

where A is the heat-transfer surface area and h_c is the coefficient of heat-transfer from the surface to the fluid, referred to as the "convective heat-transfer coefficient."

The units of the convective heat-transfer coefficient can be determined from the units of other variables:

$$Q = h_c A \Delta T$$
$$W = (h_c) m^2 K$$

Thus, the units of h_c are $W / m^2 K$.

The relationship given in Eq. (1.1) is also true for the situation where a surface is being heated due to the fluid having higher temperature than the surface. However, in this case, the direction of heat-transfer is from the fluid to the surface, and the temperature difference will now be

$$\Delta T = (\text{Mean fluid temperature}) - (\text{Surface temperature})$$

The relative temperatures of the surface and fluid determine the direction of heat-transfer and the rate at which heat-transfer take place.

As shown in Eq. (1.1), the rate of heat-transfer is not only determined by the temperature difference but also by the convective heat-transfer coefficient h_c. This is not a constant but varies quite widely depending on the properties of the fluid and the behavior of the flow. The value of h_c must depend on the thermal capacity of the fluid particle considered, that is, mC_p for the particle. The higher the density and C_p of the fluid, the better the convective heat-transfer.

Two common heat-transfer fluids are air and water, due to their widespread availability. Water is approximately 800 times more dense than air and also has a higher value of C_p. If the argument given above is valid then water has a higher thermal capacity than air and should have a better convective heat-transfer performance.

1.14.3 CONDUCTION HEAT-TRANSFER

If a fluid could be kept stationary, no convection can take place. However, it will still be possible to transfer heat by means of conduction. Conduction depends on the transfer of energy from one molecule to another within the heat-transfer medium and, in this sense, thermal conduction is analogous to electrical conduction.

Conduction can occur within both solids and fluids. The rate of heat-transfer depends on a physical property of the particular solid of fluid, termed its thermal conductivity k and the temperature gradient across the medium. The thermal conductivity is defined as the measure of the rate of heat-transfer across a unit width of material, for a unit cross-sectional area and for a unit difference in temperature.

From the definition of thermal conductivity k, it can be shown that the rate of heat-transfer is given by the relationship:

$$Q = \frac{kA\Delta T}{x}$$

ΔT is the temperature difference $T_1 - T_2$, defined by the temperature on the either side of the porous surface. The units of thermal conductivity can be determined from the units of the other variables:

$$Q = kA\Delta T / x$$
$$W = (k)m^2 K / m$$

The unit of k is $W / m^2 K / m$.

1.14.4 RADIATION HEAT-TRANSFER

The third mode of heat-transfer, radiation, does not depend on any medium for its transmission. In fact, it takes place most freely when there is a perfect vacuum between the emitter and the receiver of such energy. This is proved daily by the transfer of energy from the Sun to the Earth across the intervening space.

Radiation is a form of electromagnetic energy transmission and takes place between all matters provided that it is at a temperature above absolute zero. Infrared radiation forms only part of the overall electromagnetic spectrum. Radiation is energy emitted by the electrons vibrating in the molecules at the surface of a body. The amount of energy that can be transferred depends on the absolute temperature of the body and the radiant properties of the nanosurface.

A body that has a surface that will absorb all the radiant energy it receives is an ideal radiator, termed a "black body." Such a body will not only absorb radiation at a maximum level but will also emit radiation at a maximum level. However, in practice, bodies do not have the surface characteristics of a black body and will always absorb, or emit, radiant energy at a lower level than a black body.

It is possible to define how much of the radiant energy will be absorbed, or emitted, by a particular surface by the use of a correction factor, known as the "emissivity" and given the symbol ε. The emissivity of a surface is the measure of the actual amount of radiant energy that can be absorbed, compared to a black body. Similarly, the emissivity defines the radiant energy emitted from a surface compared to a black body. A black body will, therefore, by definition, have an emissivity ε of 1. It should be noted that the value of emissivity is influenced more by the nature of texture of clothes than its color. The practice of wearing white clothes in preference to dark clothes in order to keep cool on a hot summer's day is not necessarily valid. The amount of radiant energy absorbed is more a function of the texture of the clothes rather than the color.

Since World War II, there have been major developments in the use of microwaves for heating applications. After this time, it was realized that microwaves had the potential to provide rapid, energy-efficient heating of materials. These major applications of microwave heating today include food processing, wood drying, plastic and rubber treating as well as curing and preheating of ceramics. Broadly speaking, microwave radiation is the term associated with any electromagnetic radiation in the microwave frequency range of 300 MHz–300 Ghz. Domestic and industrial microwave ovens generally operate at a frequency of 2.45 Ghz, corresponding to a wavelength of 12.2 cm. However, not all materials can be heated rapidly by microwaves. Materials may be classified into three groups: conductors, insulators, and absorbers. Materials that absorb microwave radiation are called dielectrics; thus, microwave heating is also referred to as dielectric heating. Dielectrics have two important properties:

- They have very few charge carriers. When an external electric field is applied, there is very little change carried through the material matrix.
- The molecules or atoms comprising the dielectric exhibit a dipole movement distance. For example, the stereochemistry of covalent bonds in a water molecule, giving the water molecule a dipole movement. Water is the typical case of nonsymmetric molecule. Dipoles may be a natural feature of the dielectric or they may be induced. Distortion of the electron cloud around nonpolar molecules or atoms through the presence of an external electric field can induce a tem-

porary dipole movement. This movement generates friction inside the dielectric and the energy is dissipated subsequently as heat.

The interaction of dielectric materials with electromagnetic radiation in the microwave range results in energy absorbance. The ability of a material to absorb energy while in a microwave cavity is related to the loss tangent of the material.

This depends on the relaxation times of the molecules in the material, which, in turn, depends on the nature of the functional groups and the volume of the molecule. Generally, the dielectric properties of a material are related to temperature, moisture content, density, and material geometry.

An important characteristic of microwave heating is the phenomenon of "hot spot" formation, whereby regions of very high temperature form due to nonuniform heating. This thermal instability arises because of the nonlinear dependence of the electromagnetic and thermal properties of material on temperature. The formation of standing waves within the microwave cavity results in some regions being exposed to higher energy than others.

Microwave energy is extremely efficient in the selective heating of materials as no energy is wasted in "bulk heating" the sample. This is a clear advantage that microwave heating has over conventional methods. Microwave heating processes are currently undergoing investigation for application in a number of fields where the advantages of microwave energy may lead to significant savings in energy consumption, process time, and environmental remediation.

Compared with conventional heating techniques, microwave heating has the following additional advantages:
- higher heating rates;
- no direct contact between the heating source and the heated material;
- selective heating may be achieved;
- greater control of the heating or drying process.

1.14.5 COMBINED HEAT-TRANSFER COEFFICIENT

For most practical situations, heat-transfer relies on two, or even all three, modes occurring together. For such situations, it is inconvenient to analyze each mode separately. Therefore, it is useful to derive an overall heat-transfer coefficient that will combine the effect of each mode within a general situation. The heat-transfer in moist fabrics takes place through three modes: conduction, radiation, and distillation. With a dry fabric, only conduction and radiation occur.

1.14.6   POROSITY AND PORE SIZE DISTRIBUTION IN NANOFABRIC

The amount of porosity, that is, the volume fraction of voids within the nanofabric, determines the capacity of a nanofabric to hold water; the greater the porosity, the more water the fabric can hold. Porosity is obtained by dividing the total volume of water extruded from fabric sample by the volume of the sample:

Porosity = volume of water/volume of fabric

= (volume of water per gram sample)(density of sample)

It should be noted that most of the water is stored between the yarns rather than within them. In other words, all the water can be accommodated by the pores within the yarns, and it seems likely that the water is chiefly located there. It should be noted that pores of different sizes are distributed within a nanofabric. By a porous medium, we mean a material that containsa solid matrix with an interconnected void. The interconnectedness of the pores allows the flow of fluid through the fabric. In the simple situation (single-phase flow), the pores are saturated by a single fluid. In "two-phase flow," a liquid and a gas share the pore space.

The usual way of driving the laws governing the macroscopic variables are to begin with standard equations obeyed by the fluid and to obtain the macroscopic equations by averaging over volumes or areas containing many pores.

In defining porosity, we may assume that all the pore spaces are connected. If in fact we have to deal with a fabric in which some of the pore spaces are disconnected from the reminder, then we have to introduce an "effective porosity," defined as the ratio of the connected pore to total volume.

A further complication arises in forced convection in fabric, which is a porous medium. There may be significant thermal dispersion, that is, heat-transfer due to hydrodynamic mixing of the fluid at the pore scale. In addition to the molecular diffusion of heat, there is mixing due to the nature of the fabric.

1.14.7   THERMAL EQUILIBRIUM AND COMFORT

Some of the issues of clothing comfort most readily involve the mechanisms by which clothing materials influence heat and moisture transfer from skin to the environment. Heat flow by conduction, convection, and radiation and moisture transfer by vapor diffusion are the most important mechanisms in very cool or warm environments from the skin.

It has been recognized that the moisture-transport process in clothing under a humidity transient is one of the most important factors influencing the dynamic comfort of a wearer in practical wear situations. However, the moisture-transport process is hardly a single process as it is coupled with the heat-transfer process under dynamic conditions. Some materials will posses properties promoting rapid capillary and diffusion movement of moisture to the surface and the controlling factor will be the rate at which surface evaporation can be secured. In the initial stages of drying materials of high moisture content, it is important to obtain the highest possible rate of nanosurface evaporation. This surface evaporation is essentially the diffusion of vapor from the surface of the fabric to the surrounding atmosphere through a relatively stationary film of air in contact with its surface. This air film, in addition to presenting a resistance to the vapor flow, is itself a heat insulator. The thickness of this film rapidly decreases with increase in the velocity of the air in contact with it whilst never actually disappearing. The inner film of air in contact with the wet fabric remains saturated with vapor so long as the fabric surface has free moisture present. This results in a vapor pressure gradient through the film from the wetted solid surface to the outer air and, with large air movements, the rate of moisture diffusion through the air film will be considerable. The rate of diffusion and, hence, evaporation of the moisture will be directly proportional to the exposed area of the fabric, inversely proportional to the film thickness, and directly proportional to the inner film surface and the partial pressure of the water vapor in the surrounding air. It is of importance to note at this point that, as the layer of air film in contact with the wetted fabric undergoing drying remains saturated at the temperature of the area of contact, the temperature of the fabric surface whilst still possessing free moisture will lie very close to wet-bulb temperature of the air.

1.14.8 MOISTURE IN NANOFIBERS

The amount of moisture that a fiber can take up varies markedly. At low relative humidity, below 0.35, water is adsorbed mono-molecularly by many natural fibers. From thermodynamic reasoning, we expect the movement of water through a single fiber to occur at a rate that depends on the chemical potential gradient. Meanwhile, moisture has a profound effect on the physical properties of many fibers. Hygroscopic fibers will swell as moisture is absorbed and shrink as it is driven off. Very wet fabrics lose the moisture trapped between the threads first, and only when the threads themselves dry out will shrinkage begin. The change in volume on shrinkage is normally assumed to be linear with moisture content. With hydrophilic materials, moisture is found to reduce stiffness and increase creep, probably as a result of plasticization. Variations in moisture content can enhance creep. To describe movement of moisture at equilibrium relative humidity below unity, the idea of absorptive diffusion can be applied. Only those molecules with kinetic energies greater than the activation energy of the moisture-fiber bonds can migrate from one site to another. The driving force for absorptive diffusion is considered to be the spreading pressure, which acts over molecular surfaces in two-dimensional geometry and is similar to the vapor pressure, which acts over three-dimensional spaces.

1.14.9 CONCLUDING REMARKS

Currently, n-number of drugs are available in market, so nanotherapeutics is particularly very important to minimize health hazards caused by drugs. The most important aspect of nanotherapeutics is that it should be noninvasive and target oriented. It should have very less or no side effects. Despite using nonmaterials, one can create nanodevices for better functioning. As most of the peptides and proteins have short half-lives, frequent dosing or injections are required, which is not applicable for many situations. By using nanotherapeutics one can solve these problems. Thus, nanotherapeutics is expanding day by day in drug-delivery system.

The novel method of fixation of proteins on MNPs proposed in the work was successfully realized on the example of albumin and thrombin. The blood plasma proteins are characterized by a high biocompatibility and allow decreasing toxicity of nanoparticles administered into organism.

The method is based on the ability of proteins to form inter chain co-valent bonds under the action of free radicals. The reaction mixture for stable coatings obtaining should consist on protein solution, nanoparticles containing metals of variable valence (e.g., Fe, Cu, Cr) and water-soluble initiator of free radicals generation. In this work, albumin and thrombin were used for coating being formed on magnetite nanoparticles. Hydrogen peroxide served as initiator. By the set of physical (ESR-spectroscopy, ferromagnetic resonance, dynamic and Rayleigh light scattering, IR-spectroscopy) and biochemical methods, it was proved that the coatings obtained are stable and formed on individual nanoparticles because free radical processes are localized strictly in the adsorption layer. The free radical linking of thrombin on the surface of nanoparticles has been shown to almost completely keep native properties of the protein molecules. As the method provides enzyme activity and formation of thin stable protein layers on individual nanopaticles, it can be successfully used for various biomedical goals concerning a smart delivery of therapeutic products and biologically active substances (including enzymes). It reveals principally novel technologies of one-step creation of biocompatible MNSs with mul-tiprotein polyfunctional coatings that meet all the requirements and con-tain both biovectors and therapeutic products (**Figure 1.13**).

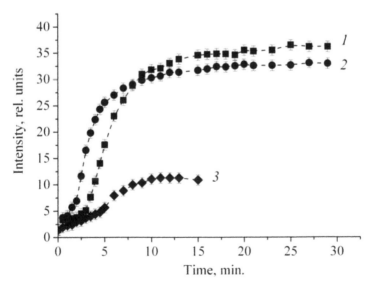

FIGURE 1.13   The scheme of magnetically targeted nanosystem for a smart delivery of therapeutic products based on the free radical protein cross-linking.

For heat flow analysis of wet porous nanosystems, the liquid is water and the gas is air. Evaporation or condensation occurs at the interface between the water and air so that the air is mixed with water vapor. A flow of the mixture of air and vapor may be caused by external forces, for instance, by an imposed pressure difference. The vapor will also move relative to the gas by diffusion from regions where the partial pressure of the vapor is higher to those where it is lower.

Heat flow in porous nanosystems is the study of energy movement in the form of heat, which occurs in various types of processes. The transfer of heat in porous nanostructure fabrics occurs from the high- to the low-temperature regions. Therefore, a temperature gradient has to exist between the two regions for heat-transfer to happen. It can be done by conduction (within one porous solid or between two porous solids in contact), by convection (between two fluids or a fluid and a porous solid in direct contact with the fluid), by radiation (transmission by electromagnetic waves through space), or by combination of the above three methods.

KEYWORDS

- **Nanosystems**
- **Nanotechnology**
- **Overview**

REFERENCES

1. Wu, Y.; and Clark, R. L.; *J. Colloid Interface Sci.* **2007,** *310(2),* 529.
2. Jaworek, A.; and Sobczyk, A. T.; *J. Electrostat.* **2008,** *66(3–4),* 197.
3. Koombhongse, S.; Liu, W.; and Reneker, D. H.; *J. Polym. Sci. Part B: Polym. Phys.* **2001,** *39(21),* 2598.
4. Yu, Q. Z.; Li, Y.; Wang, M.; and Chen, H. Z.; *Chin. Chem. Lett.* **2008,** *19(2),* 223.
5. Luo, C. J.; Nangrejo, M.; and Edirisinghe, M.; *Polymer.* **2010,** *51(7),* 1654.
6. Zheng, J.; He, A.; Li, J.; Xu, J.; and Han, C. C.; *Polymer.* **2006,** *47(20),* 7095.
7. Leong, M. F.; Chian, K. S.; Mhaisalkar, P. S.; Ong, W. F.; and Ratner, B. D.; *J. Biomed. Mater. Res. Part A.* **2009,** *89(4),* 1040.
8. Servoli, E.; Ruffo, G. A.; and Migliaresi, C.; *Polymer.* **2010,** *51(11),* 2337.
9. Dayal, P.; and Kyu, T.; *J. Appl. Phys.* **2006,** *100(4),* 043512.
10. Megelski, S.; Stephens, J. S.; Chase, D. B.; and Rabolt, J. F.; *Macromolecules.* **2002,** *35(22),* 8456.

11. Rangkupan, R.; and Reneker, D. H.; *J. Metal. Mater. Miner.* **2003,** *12(2),* 81.
12. Dalton, P. D.; Klinkhammer, K.; Salber, J.; Klee, D.; and Möller, M.; *Biomacromolecules.* **2006,** *7(3),* 686.
13. Shin, J. W.; et al. In: Proceedings of the 3rd Kuala Lumpur International Conference on Biomedical Engineering (IFMBE). Kuala Lumpur, Malaysia; **2006,** 692 p.
14. Bonani, W.; Maniglio, D.; Motta, A.; Tan, W.; and Migliaresi, C.; *J. Biomed. Mater. Res. Part B: Appl. Biomater.* **2011,** *96(2),* 276.
15. Lin, T.; Wang, H.; and Wang, X.; *Adv. Mater.* **2005,** *17(22),* 2699.
16. Gupta, P.; and Wilkes, G. L.; *Polymer.* **2003,** *44(20),* 6353.
17. Jiang, H.; Hu, Y.; Li, Y.; Zhao, P.; Zhu, K.; and Chen, W.; *J. Controlled Release.* **2005,** *108(2–3),* 237.
18. Díaz, J. E.; Fernández-Nieves, A.; Barrero, A.; Márquez, M.; and Loscertales, I. G.; *J. Phys.: Conf. Ser.* **2008,** *127(1),* 012008.63 *Adv. Elect. Setups Spec. Fibre Mesh Morphol.*
19. Stankus, J. J.; Guan, J.; Fujimoto, K.; and Wagner, W. R.; *Biomaterial.* **2006,** *27(5),* 735.
20. Brauker, J. H.; Carr-Brendel, V. E.; Martinson, L. A.; Crudele, J.; Johnston, W. D.; and Johnson, R. C.; *J. Biomed. Mater. Res.* **1995,** *29(12),* 1517.
21. Lu, B.; et al. *Small.* **2010,** *6(15),* 1612.
22. Fang, D.; Hsiao, B. S.; and Chu, B.; *Polym. Preprints.* **2003,** *44(2),* 59.
23. Zhou, F. -L.; Gong, R. -H.; and Porat, I.; *J. Appl. Polym. Sci.* **2010,** *115(5),* 2591.
24. Theron, S. A.; Yarin, A. L.; Zussman, E.; and Kroll, E.; *Polymer.* **2005,** *46(9),* 2889.
25. Varesano, A.; Carletto, R. A.; and Mazzuchetti, G.; *J. Mater. Proc. Technol.* **2009,** *209(11),* 5178.
26. Kim, G.; Cho, Y. -S.; and Kim, W. D.; *Euro. Polym. J.* **2006,** *42(9),* 2031.
27. Yamashita, Y.; Ko, F.; Miyake, H.; and Higashiyama, A.; *Sen'iGakkaishi.* **2008,** *64(1),* 24.61 *Adv. Elec. Setups Spec. Fibre Mesh Morphol.*
28. Zhou, F. -L.; Gong, R. -H.; and Porat, I.; *Polym. Eng. Sci.* **2009,** *49(12),* 2475.
29. Kumar, A.; Wei, M.; Barry, C.; Chen, J.; and Mead, J.; *Macromol. Mater. Eng.* **2010,** *295(8),* 701.
30. Dosunmu, O. O.; Chase, G. G.; Kataphinan, W.; and Reneker, D. H.; *Nanotechnol.* **2006,** *17(4),* 1123.
31. Badrossamay, M. R.; McIlwee, H. A.; Goss, J. A.; and Parker, K. K.; *Nano Lett.* **2010,** *10(6),* 2257.
32. Srivastava, Y.; Marquez, M.; and Thorsen, T.; *J. Appl. Polym. Sci.* **2007,** *106(5),* 3171.
33. Jirsak, O.; Sanetrnik, F.; Lukas, D.; Kotek, V.; Martinova, L.; and Chaloupek, J.; Inventors; Technicka Universita V Liberci, Assignee; US Patent 7585437B2. 2009.
34. Lukas, D.; Sarkar, A.; and Pokorny, P.; *J. Appl. Phys.* **2008,** *103(8),* 084309.
35. Yarin, A.; and Zussman, E.; *Polymer.* **2004,** *45(9),* 2977.
36. Thoppey, N. M.; Bochinski, J. R.; Clarke, L. I.; and Gorga, R. E.; *Polymer.* **2010,** *51(21),* 4928.
37. Tang, S.; Zeng, Y.; and Wang, X.; *Polym. Eng. Sci.* **2010,** *50(11),* 2252.
38. Cengiz, F.; Dao, T. A.; and Jirsak, O.; *Polym. Eng. Sci.* **2010,** *50(5),* 936.
39. Jirsak, O.; Sysel, P.; Sanetrnik, F.; Hruza, J.; and Chaloupek, J.; *J. Nanomater.* **2010,** *2010,* 1.
40. Wang, X.; Niu, H.; Lin, T.; and Wang, X.; *Polym. Eng. Sci.* **2009,** *49(8),* 1582.

41. Liu, Y.; He, J. -H.; and Yu, J. -Y.; *J. Phys.: Conf. Ser.* **2008,** *96,* 012001.
42. Liu, Y.; Dong, L.; Fan, J.; Wang, R.; and Yu, J. -Y.; *J. Appl. Polym. Sci.* **2011,** *120(1),* 592.62 *Elec. Adv. Biomed. Appl. Therapies.*
43. Salem, D. R.; In: Nanofibers and Nanotechnology in Textiles. Eds. Brown, P. J.; and Stevens, K.; Boca Raton, FL, USA: CRC Press; **2007,** 1 p.
44. Kelly, A. J.; *J. Aerosol Sci.* **1994,** *25(6),* 1159.
45. Sun, D.; Chang, C.; Li, S.; and Lin, L.; *Nano Lett.* **2006,** *6(4),* 839.
46. Chang, C.; Limkrailassiri, K.; and Lin, L.; *Appl. Phys. Lett.* **2008,** *93(12),* 123111.
47. Levit, N.; and Tepper, G.; *The J. Supercrit. Fluids.* **2004,** *31(3),* 329.
48. Larrondo, L.; and Manley, R. S. J.; *J. Polym. Sci. Part B. Polym. Phys.* **1981,** *19(6),* 909.
49. Liu, Y.; Wang, X.; Li, H.; Yan, H.; and Yang, W.; *Society Plast. Eng. Plast. Res. Online.* **2010,** 10.1002/spepro.003055.
50. Dalton, P. D.; Grafahrend, D.; Klinkhammer, K.; Klee, D.; and Möller, M.; *Polymer.* **2007,** *48(23),* 6823.
51. Volpato, F. Z.; Ramos, S. L. F.; Motta, A.; and Migliaresi, C.; *J. Bioactive Compat. Polym.* **2011,** *26(1),* 35.
52. Deitzel, J.; Kleinmeyer, J. D.; Hirvonen, J. K.; and Beck Tan, N. C.; *Polymer.* **2001,** *42(19),* 8163.
53. Li, D.; Wang, Y.; and Xia, Y.; *Nano Lett.* **2003,** *3(8),* 1167.
54. Huang, Z. -M.; Zhang, Y. -Z.; Kotaki, M.; and Ramakrishna, S.; *Compos. Sci. Technol.* **2003,** *63(15),* 2223.
55. Teo, W. E.; Kotaki, M; Mo, X. M.; and Ramakrishna, S.; *Nanotechnology.* **2005,** *16(6),* 918.
56. Theron, A.; Zussman, E.; and Yarin, A. L.; *Nanotechnology.* **2001,** *12(3),* 384.
57. Li, D.; Wang, Y.; and Xia, Y.; *Adv. Mater.* **2004,** *16(4),* 361.
58. Katta, P.; Alessandro, M.; Ramsier, R. D.; and Chase, G. G.; *Nano Lett.* **2004,** *4(11),* 2215.
59. Hong, J. K.; and Madihally, S. V.; *Tissue Eng.* **2011,** *17(2),* 125.
60. Thorvaldsson, A.; Engström, J.; Gatenholm, P.; and Walkenström, P.; *J. Appl. Polym. Sci.* **2010,** *118(1),* 511.
61. Hou, Q.; Grijpma, D. W.; and Feijen, J.; *Biomaterial.* **2003,** *24(11),* 1937.
62. Nam, J.; Huang, Y.; Agarwal, S.; and Lannutti, J.; *Tissue Eng.* **2007,** *13(9),* 2249.
63. Kim, T. G.; Chung, H. J.; and Park, T. G.; *Acta Biomater.* **2008,** *4(6),* 1611.
64. Baker, B. M.; Gee, A. O.; Metter, R. B.; Nathan, A. S.; Marklein, R. A.; Burdick, J. A.; and Mauck, R. L.; *Biomater.* **2008,** *29(15),* 2348.
65. Whited, B. M.; Whitney, J. R.; Hofmann, M. C.; Xu, Y.; and Rylander, M. N.; *Biomater.* **2011,** *32(9),* 2294.
66. Gentsch, R.; Boysen, B.; Lankenau, A.; and Börner, H. G.; *Macromol. Rapid Comm.* **2010,** *31(1),* 59.
67. Soliman, S.; et al. *Acta Biomater.* **2010,** *6(4),* 1227.
68. Santos, M. I.; et al. *Biomaterial.* **2008,** *29(32),* 4306.
69. Tuzlakoglu, K.; Bolgen, N.; Salgado, A. J.; Gomes, M. E.; Piskin, E.; and Reis, R. L.; *J. Mater. Sci.: Mater. Med.* **2005,** *16(12),* 1099.65 *Adv. Elec. Setups Spec. Fibre Mesh Morphol.*

70. Thorvaldsson, A.; Stenhamre, H.; Gatenholm, P.; and Walkenström, P.; *Biomacromol.* **2008,** *9(3),* 1044.
71. Smit, E.; Büttner, U.; and Sanderson, R. D.; *Polymer.* **2005,** *46(8),* 2419.
72. Khil, M. -S.; Bhattarai, S. R.; Kim, H. -Y.; Kim, S. -Z.; and Lee, K. -H.; *J. Biomed. Mater. Res. Part B. Appl. Biomater.* **2005,** *72(1),* 117.
73. Teo, W. E.; and Ramakrishna, S.; *Nanotechnology.* **2006,** *17(14),* R89.64 *Elec. Adv. Biomed. Appl. Therapies.*
74. Simonet, M.; Schneider, O. D.; Neuenschwander, P.; and Stark, W. J.;*Polym. Eng. Sci. Eng.* **2007,** *47(12),* 2020.
75. Leong, M. F.; Rasheed, M. Z.; Lim, T. C.; and Chian, K. S.; *J. Biomed. Mater. Res. Part A.* **2009,** *91(1),* 231.
76. Schneider, O. D.; Weber, F.; Brunner, T. J.; Loher, S.; Ehrbar, M.; Schmidlin, P. R.; and Stark, W. J.; *Acta Biomater.* **2009,** *5(5),* 1775.
77. Nam, Y. S.; Yoon, J. J.; and Park, T. G.; *J. Biomed. Mater. Res.* **2000,** *53(1),* 1.
78. Lee, Y. H.; et al. *Biomaterial.* **2005,** *26(16),* 3165.
79. Curtis, A. S. G.; Gadegaard, N.; Dalby, M. J.; Riehle, M. O.; Wilkinson, C. D. W.; and Aitchison, G.; *IEEE Transact. Nanobiosci.* **2004,** *3(1),* 61.
80. Wang, Y.; et al. *Biomaterial.* **2010,** *31(17),* 4672.
81. Kim, G.; Son, J.; Park, S.; and Kim, W.; *Macromol. Rapid Comm.* **2008,** *29(19),* 1577.
82. Ahn, S. H.; Koh, Y. H.; and Kim, G. H.; *J. Micromech. Microeng.* **2010,** *20(6),* 065015.
83. Yoon, H.; and Kim, G.; *J. Pharmaceut. Sci.* **2011,** *100(2),* 424.
84. Tayalia, P.; Mendonca, C. R.; Baldacchini, T.; Mooney, D. J.; and Mazur, E.; *Adv. Mater.* **2008,** *20(23),* 4494.
85. Sachlos, E.; and Czernuszka, J. T.; *Euro. Cells Mater.* **2003,** *5,* 29.
86. Mota, C.; Puppi, D.; Dinucci, D.; Errico, C.; Bártolo, P.; and Chiellini, F.; *Mater.* **2011,** *4(3),* 527.
87. Wang, L; Pai, C. -L.; Boyce, M. C.; and Rutledge, G. C.; *Appl. Phys. Lett.* **2009,** *94(15),* 151916.
88. Casper, C. L.; Stephens, J. S.; Tassi, N. G.; Chase, D. B.; and Rabolt, J. F.; *Macromol.* **2004,** *37(2),* 573.
89. Rosso, F.; Giordano, A.; Barbarisi, M.; and Barbarisi, A.; *J. Cell. Physiol.* **2004,** *199(2),* 174.
90. Boyan, B. D.; Lossdörfer, S.; Wang, L.; Zhao, G.; Lohmann, C. H.; Cochran, D. L.; and Schwartz, Z.; *Euro. Cells Mater.* **2003,** *6,* 22.
91. Chung, T. -W.; Liu, D. -Z.; Wang, S. -Y.; and Wang, S. -S.; *Biomaterial.* **2003,** *24(25),* 4655.
92. Moroni, L.; Licht, R.; de Boer, J.; de Wijn, J. R.; and van Blitterswijk, C. A.; *Biomaterial.* **2006,** *27(28),* 4911.
93. Truong, Y. B.; et al. *Biomed. Mater.* **2010,** *5(2),* 25005.
94. Lee, K. Y.; and Yuk, S. H.; *Progress Polym. Sci.* **2007,** *32(7),* 669.
95. Tao, S. L.; and Desai, T. A.; *Adv. Drug Delive. Rev.* **2003,** *55(3),* 315.66 *Elec. Adv. Biomed. Appl. Therapies.*
96. Moroni, L.; de Wijn, J. R.; and van Blitterswijk, C. A.; *J. Biomater. Sci. Polym. Ed.* **2008,** *19(5),* 543.
97. Luginbuehl, V.; Meinel, L.; Merkle, H. P.; and Gander, B.; *Euro. J. Pharmaceut. Biopharmaceut.* **2004,** *58(2),* 197.

98. Willerth, S. M.; and Sakiyama-Elbert, S. E.; *Adv. Drug Delive. Rev.* **2007,** *59(4–5),* 325.

99. Biondi, M.; Ungaro, F.; Quaglia, F.; and Netti, P. A.; *Adv. Drug Delive. Rev.* **2008,** *60(2),* 229.

100. Munir, M. M.; Suryamas, A. B.; Iskandar, F.; and Okuyama, K.; *Polymer.* **2009,** *50(20),* 4935.

101. Qi, Z.; Yu, H.; Chen, Y.; and Zhu, M.; *Mater. Lett.* **2009,** *63(3–4),* 415.

102. Luna-Bárcenas, G.; Kanakia, S. K.; Sanchez, I. C.; and Johnston, K. P.; *Polym.* **1995,** *36(16),* 3173.67 *Adv. Elec. Setups Spec. Fibre Mesh Morphol.*

103. Liu, J.; Shen, Z.; Lee, S. -H.; Marquez, M.; and McHugh, M. A.; *The J. Supercrit. Fluids.* **2010,** *53(1–3),* 142.

104. McCann, J. T.; Marquez, M.; and Xia, Y.; *J. Am. Chem. Soc.* **2006,** *128(5),* 1436.

105. You, Y.; Youk, J. H.; Lee, S. W.; Min, B. -M.; Lee, S. J.; and Park, W. H.; *Mater. Lett.* **2006,** *60(6),* 757.

106. Lyoo, W. S.; Youk, J. H.; Lee, S. W.; and Park, W. H.; *Mater. Lett.* **2005,** *59(28),* 3558.

107. Chakraborty, S. Liao, I. -C.; Adler, A.; and Leong, K. W.; *Adv. Drug Delive. Rev.* **2009,** *61(12),* 1043.

108. Srikar, R.; Yarin, A. L.; Megaridis, C. M.; Bazilevsky, A. V.; and Kelley, E.; *Langmuir.* **2008,** *24(3),* 965.

109. Gandhi, M.; Srikar, R.; Yarin, A. L.; Megaridis, C. M.; and Gemeinhart, R. A.; *Mole. Pharmaceut.* **2009,** *6(2),* 641.

110. Wang, M.; Yu, J. H.; Kaplan, D. L.; and Rutledge, G. C.; *Macromole.* **2006,** *39(3),* 1102.

111. Gupta, A. K.; Gupta, M.; *Biomaterial.* **2005,** *26,* 3995

112. Vatta, L. L.; Sanderson, D. R.; Koch, K. R.; *Pure Appl. Chem.* **2006,** *78,* 1793.

113. Laurent, S.; et al. *Chem. Rev.* **2008,** *108,* 2064.

CHAPTER 2

A CASE STUDY ON PERFORMANCE OF NANOPARTICLE AGGREGATES IN POLYMERIC NANOCOMPOSITES

G. E. ZAIKOV and A. K. HAGHI

CONTENTS

2.1   INTRODUCTION

Nanoparticle structure is defined by chemical interactions between atoms nature forming them. The fundamental properties of nanoparticles, forming in strongly nonequilibrium conditions is their ability to:

  (i)   Structures self-organization by adaptation to external influence;
 (ii)   An optimal structure self-choice in bifurcation points, corresponding to preceding structure stability threshold and new stable formation;
(iii)   A self-operating synthesis (self-assembly) of stable nanoparticles, which is ensured by information exchange about system structural state in the previous bifurcation point at a stable structure self-choice in the following beyond it bifurcation point.

These theoretical postulates were confirmed experimentally. In particular, it has been shown that nanoparticle sizes are not arbitrary ones, but change discretely and obey to synergetics laws. This postulate is important from the practical point of view, since nanoparticle size is the information parameter, defining surface energy critical level [1–15].

Let us consider these general definitions in respect to polymer particulate filled nanocomposites, for which there are certain distinctions with the considered above criterions. As it is well-known, nanofiller aggregation processes in either form are inherent in all types of polymer nanocomposites and influence essentially on their properties. In this case, although nanofiller initial particles have size (diameter) less than 100 nm, but these nanoparticle aggregates can exceed essentially the indicated above boundary value for nanoworld objects. Secondly, nanofiller particles aggregates are formed at the expence of physical interactions, but not chemical ones. Therefore the present update reports the synergetics laws applicability for nanofiller aggregation processes and interfacial phenomena description in particulate-filled polymer nanocomposites on the example of nanocomposite polypropylene/calcium carbonate [16–37].

2.2   EXPERIMENTAL

Polypropylene (PP) with average molecular weight M_w was used as matrix polymer and nanodimensional calcium carbonate ($CaCO_3$) in the form of compound was used as nanofiller.

Nanocomposites PP/CaCO$_3$ was prepared by components mixing in melt on twin-screw extruder. Mixing was performed at temperature 463–503 K and screw speed of 50 rpm during 5 min. Testing samples were obtained by casting under pressure method on casting machine at temperature of 483 K and pressure of 43 MPa.

The electron microscopy was used for nanocomposites PP/CaCO$_3$ structure study. Study objects were prepared in liquid nitrogen with the purpose of microscopic sections obtaining. The scanning electron microscope with autoemissive cathode of high resolution was used for microscopic sections surface images obtaining. Images were obtained in the mode of low-energetic secondary electrons, since this mode ensures the highest resolution.

Uniaxial tension mechanical tests have been performed on the samples. The tests have been conducted on universal testing apparatus at temperature of 293 K and strain rate ~2 10^{-3} s^{-1}.

2.3   RESULTS AND DISCUSSION

A particulate nanofiller particles aggregate size (diameter) D_{agr} estimation can be performed according to the following formula:

$$k(r)\lambda = \left[\left(\frac{25.1\pi D_{agr}^{1/3}}{W_n}\right)^{1/3} - 2\right]\frac{D_{agr}}{2} \tag{2.1}$$

where $k(r)$ is an aggregation parameter, λ is distance between nanofiller particles, W_n is nanofiller mass contents in mass per cent.

In its turn, the value $k(r)\lambda$ is determined within the frameworks of strength dispersion theory with the help of the relationship:

$$\tau_n = \tau_m + \frac{Gb_B}{k(r)\lambda} \tag{2.2}$$

where τ_n and τ_m are yield stress in compression testing of nanocomposite and matrix polymer, accordingly, G is shear modulus, b_B is Burders vector.

The general relationship between normal stress σ and shear stress τ has the following look:

$$\tau = \frac{\sigma}{\sqrt{3}} \tag{2.3}$$

Young's modulus E and shear modulus G are connected between themselves by the simple relationship:

$$G = \frac{E}{d_f} \tag{2.4}$$

where d_f is nanocomposite structure fractal dimension, which is determined according to the equation:

$$d_f = (d-1)(1+v) \tag{2.5}$$

where d is the dimension of Euclidean space, in which fractal is considered, v is Poisson's ratio, which is estimated by the mechanical testing results with the help of the relationship:

$$\frac{\sigma_Y}{E_n} = \frac{1-2v}{6(1+v)} \tag{2.6}$$

where σ_Y and E_n are yield stress and elasticity modulus of nanocomposite, respectively.

Burgers vector value b_B for polymeric materials is determined according to the equation [11]:

$$b_B = \left(\frac{60.5}{C_\infty}\right)^{1/2} \tag{2.7}$$

where C is characteristic ratio, connected with d_f by the equation:

$$C_\infty = \frac{2d_f}{d(d-1)(d-d_f)} + \frac{4}{3} \tag{2.8}$$

The calculation according to the **Eqs. (2.1–2.8)** showed $CaCO_3$ nanoparticle aggregates mean diameter growth from 85 up to 190 nm within the range of $W_n = 1 - 7$ mass per cent for the considered nanocomposites $PP/CaCO_3$. These calculations can be confirmed experimentally by the electron microscopy methods. In Figure 2.1 the nanocomposites PP/CaCO$_3$ with nanofiller contents $W_n = 1$ and 4 mass per cent sections electron micrographs are adduced. As one can see, if at $W_n = 1$ mass per cent nanofiller particles are not aggregated practically, that is, their diameter is close to CaCO$_3$ initial nanoparticle diameter (~80 nm), then at $W_n = 4$ mass per cent the initial nanoparticle aggregation is observed even visually and these particles aggregates sizes are varied within the limits of 80–360 nm.

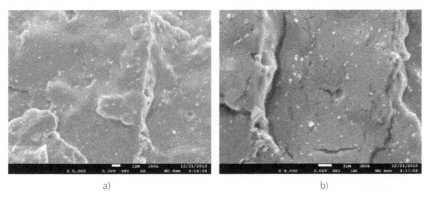

a)   b)

FIGURE 2.1 Electron micrographs of sections of nanocomposites PP/CaCO$_3$ with nanofiller mass contents $W_n = 1$ (a) and 4 (b) mass per cent.

The adduced above estimations correspond to the results of calculation according to the **Eq. (2.1)** at the indicated CaCO$_3$ contents: 85 and 142 nm, accordingly. Hence, the considered above technique gives reliable enough estimations of nanofiller particles aggregates diameter.

It has been shown earlier on the example of different physical-chemical processes, that the self-similarity function has an iteration type function look, connecting structural bifurcation points by the relationship:

$$A_m = \frac{\lambda_n}{\lambda_{n+1}} = \Delta_i^{1/m} \tag{2.9}$$

Where A_m is the measure of aggregate structure adaptability to external influence, λ_n and λ_{n+1} are preceding and subsequent critical values of operating parameter at the transition from preceding to subsequent bifurcation point, Δ_i is the structure stability measure, remaining constant at its re-organization up to symmetry violation, m is an exponent of feedback type; the value $m = 1$ corresponds to linear feedback, at which transitions on other spatial levels are realized by multiplicative structure reproduction mechanism and at $m \geq 2$ (nonlinear feedback)—replicative (with structure improvement) one.

Selecting as the operating parameter critical value nanoparticle aggregates diameter D_{agr} at successive W_n change, the dependence of adaptability measure A_m on W_n can be plotted, which is shown in Figure 2.2. As one can see, for the considered nanocomposites the condition is fulfilled

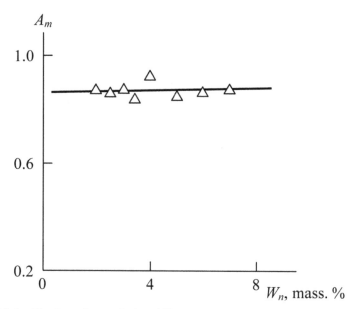

FIGURE 2.2   The dependence of adaptability measure A_m on nanofiller mass contents W_n for nanocomposites PP/CaCO$_3$.

$$A_m = \frac{D_{agri}}{D_{agri+1}} = \text{const} = 0.899 \qquad (2.10)$$

With precision of 2 per cent. This means, that aggregation processes in the considered nanocomposites obey to the synergetics laws, although their aggregates size exceeds the boundary value of 100 nm for nanoworld [37]. Let us note the important aspect of the dependence $A_m(W_n)$, adduced in Figure 2.2. The condition $A_m = \text{const}$ is kept irrespective of gradation, with which W_n changes −0.5 or 1.0 mass per cent.

As it is known, an interfacial layer in polymer nanocomposites can be considered as a result of two fractal objects (polymer matrix and nano-filler particles surface) interaction, for which there is the only linear scale l, defining these objects interpenetration distance. Since the filler elasticity modulus is, as a rule, considerably higher than the corresponding parameter for polymer matrix, then the indicated interaction comes to filler indentation in polymer matrix and then $l = l_{if}$, where l_{if} is interfacial layer thickness . In this case it can be written:

$$l_{if} = a\left(\frac{D_{agr}}{2a}\right)^{2(d-d_{surf})/d}$$ (2.11)

Where a is a lower linear scale of fractal behavior, which for polymeric materials is accepted equal to statistical segment length l_{st}, d_{surf} is nano-filler particles (aggregates of particles) surface fractal dimension.
   l_{st} is determined according to the equation:

$$l_{st} = l_0 C_\infty$$ (2.12)

Where l_0 is the main chain skeletal bond length, equal to 0.154 nm for PP.
   The dimension d_{surf} is calculated in the following succession. First the nanofiller particles aggregate density ρ_n is estimated according to the formula:

$$\rho_n = 188\left(D_{agr}\right)^{1/3}$$ (2.13)

Then the indicated aggregate specific surface S_u is determined:

$$S_u = \frac{6}{\rho_n D_{agr}}$$ (2.14)

And at last, the value d_{surf} calculation can be fulfilled with the help of the equation:

$$S_u = 410\left(\frac{D_{agr}}{2}\right)^{d_{surf}-d}$$ (2.15)

where S_u is given in m²/g, D_{agr} —in nm.

The calculation according to the offered technique has shown l_{if} increase from 1.78 up to 5.23 nm at W_n enhancement within the range of 1–7 mass per cent. The estimations according to the **Eq. (2.9)**, where as λ_n and λ_{n+1} the values $l_{if\,n}$ and $l_{if\,n+1}$ were accepted, showed that the following condition was fulfilled:

$$A_m = \frac{l_{if\,n}}{l_{if\,n+1}} = 0.880 \qquad (2.16)$$

With the precision of 7 per cent. Hence, an interfacial layers formation in polymer nanocomposites, characterizing interfacial phenomena in these nanomaterials, obeys to the synergetics laws with the same adaptability measure, as nanofiller particles aggregation. Nevertheless, it should be noted, that this analogy is not complete for the considered nanocomposites within the range of W_n = 1–7 mass per cent the value D_{agr} increases in 2.24 times, whereas the value l_{if} does almost in three times.

Let us consider further the exponent m in the Eq. (2.9) choice, characterizing feedback type in aggregation process. As it has been noted above, this exponent is equal to 2 in case of aggregates structure improvement, which can be characterized by their fractal dimension d_f^{agr}. This dimension can be calculated with the help of the equation:

$$\rho_n = \rho_{dens}\left(\frac{D_{agr}}{2a}\right)^{d_f^{agr}-d} \qquad (2.17)$$

where ρ_{dens} is massive material density, which is equal to 2,000 kg/m³ for $CaCO_3$, a is a lower linear scale of fractal behavior, accepted equal to 10 nm.

In Figure 2.3 the dependence of d_f^{agr} on $CaCO_3$ mass contents W_n for the considered nanocomposites is adduced. As one can see, within the studied range of W_n the essential d_f^{agr} growth (from 2.34 up to 2.73 at general d_f^{agr} variation within the limit of 2.0 to 2.95) is observed, that can be classified as nanofiller particles aggregates structure improvement, as a minimum, by two reasons: their disaggregation level reduction and critical structural defect sizes decreasing. Therefore, proceeding from the said above, it should be accepted that $m = 2$, which according to the Eq. (2.9) gives $\Delta_I = 0.808$. Let us note, that this Δ_I value defines very stable nanostructures. So, for self-operating nano-solid solutions synthesis the values $\Delta_I = 0.255 - 0.465$ at $m = 2$ were obtained and in addition it has been shown that an optimal technological regime indicator is $\Delta_i = 0.465$ attainment at nonlinear feedback ($m = 2$) realization.

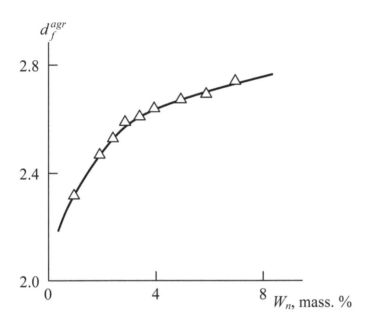

FIGURE 2.3 The dependence of nanofiller particles aggregates structure fractal dimension d_f^{agr} on its mass contents W_n for nanocomposites $PP/CaCO_3$.

Let us consider in conclusion the possibility of $CaCO_3$ nanoparticle aggregation process prediction within the frameworks of synergetics treatment. In Figure 2.4 the comparison of $CaCO_3$ aggregates diameter values, calculated according to the Eqs. (2.1) D_{agr} and (2.9) D_{agr}^{syn} at $A_m = const = 0.899$. As one can see, this comparison demonstrates very good conformity of nanofiller particles aggregates diameter, calculated by both indicated methods (the average discrepancy of D_{agr} and D_{agr}^{syn} makes up 2%).

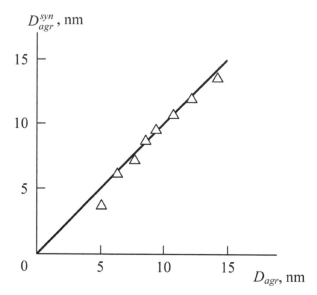

FIGURE 2.4   The comparison of nanofiller particles aggregates diameter, calculated according to the Eq. (2.1) D_{agr} and Eq. (2.9) D_{agr}^{syn}, for nanocomposites PP/CaCO$_3$.

CONCLUDING REMARKS

Hence, the present papers results have demonstrated that disperse nanoparticle aggregates formation in polymer nanocomposites obeys to synergetics laws even at these aggregates size larger than upper dimensional boundary for nanoparticles, equal to 100 nm. These aggregates are formed by structure reproduction replicative mechanism with nonlinear feedback. In this case nanofiller contents discrete change points are bifurcation points. The synergetics methods can be used for prediction of forming nanoparticle aggregates size. The parameters, characterizing interfacial phenomena in polymer nanocomposites (for example, interfacial layer thickness) obey also to synergetics laws.

KEYWORDS

- **Nanoparticles**
- **Polypropylene/calcium carbonate**

REFERENCES

1. Ziabari, M.; Mottaghitalab, V.; McGovern, S. T.; and Haghi, A. K.; *Chem. Phys. Lett.* **2008a,** *25,* 3071.
2. Ziabari, M.; Mottaghitalab, V.; McGovern, S. T.; and Haghi, A. K.; *Nanoscale Res. Lett.* **2007,** *2,* 297.
3. Ziabari, M.; Mottaghitalab, V.; and Haghi, A. K.; *Korean J. Chem. Eng.* **2008b,** *25,* 905.
4. Ziabari, M.; Mottaghitalab, V.; and Haghi, A. K.; *Korean J. Chem. Eng.* **2008c,** *25,* 919.
5. Haghi, A. K.; and Akbari, M.; *Phys. Status Solidi.* **2007,** *204,* 1830.
6. Kanafchian, M.; Valizadeh, M.; and Haghi, A. K.; *Korean J. Chem. Eng.* **2011a,** *28,* 428.
7. Kanafchian, M.; Valizadeh, M.; and Haghi, A. K.; *Korean J. Chem. Eng.* **2011b,** *28,* 763.
8. Kanafchian, M.; Valizadeh, M.; and Haghi, A. K.; *Korean J. Chem. Eng.* **2011c,** *28,* 751.
9. Kanafchian, M.; Valizadeh, M.; and Haghi, A. K.; *Korean J. Chem. Eng.* **2011d,** *28,* 445.
10. Afzali, A.; Mottaghitalab, V.; Motlagh, M.; and Haghi, A. K.; *Korean J. Chem. Eng.* **2010,** *27,* 1145.
11. Wan, Y. Q.; Guo, Q.; and Pan, N.; *Int. J. Nonlinear Sci. Numerical Simul.* **2004,** *5,* 5.
12. Feng, J. J.; *J. Non-Newtonian Fluid Mech.* **2003,** *116,* 55.
13. He, J.; Wan, Y.; and Yu, J. -Y.; *Polym.* **2005,** *46,* 2799.
14. Zussman, E.; Theron, A.; and Yarin, A. L.; *Appl. Phys. Lett.* **2003,** *82,* 73.
15. Reneker, D. H.; Yarin, A. L.; Fong, H.; and Koombhongse, S.; *J. Appl. Phys.* **2000,** *87,* 4531.
16. Theron, S. A.; Yarin, A. L.; Zussman, E.; and Kroll, E.; *Polym.* **2005,** *46,* 2889.
17. Huang, Z. -M.; Zhang, Y. -Z.; Kotak, M.; and Ramakrishna, S.; *Compos. Sci. Technol.* **2003,** *63,* 2223.
18. Schreuder-Gibson, H. L.; et al. *J. Adv. Mater.* **2002,** *34(3),* 44.
19. Ma, Z.; Kotaki, M.; Inai, R.; and Ramakrishna, S.; *Tissue Eng.* **2005,** *11,* 101.
20. Ma, Z.; Kotaki, M.; Yong, T.; He, W.; and Ramakrishna, S.; *Biomater.* **2005,** *26,* 2527.
21. Jin, H. J.; Fridrikh, S.; Rutledge, G. C.; and Kaplan, D.; *Abst. Papers Am. Chem. Soc.* **2002,** *224(1/2),* 408.
22. Luu, Y. K.; Kim, K.; Hsiao, B. S.; Chu, B.; and Hadjiargyrou, M.; *J. Contr. Rel.* **2003,** *89,* 341.
23. Senecal, K. J.; Samuelson, L.; Sennett, M.; Schreuderm, Inventors US 0045547. **2001.**
24. Sawicka, K.; Goum, P.; and Simon, S.; *Sensors and Actuators B.* **2005,** *108,* 585.
25. Fujihara, K.; Kotak, M.; and Ramakrishn, S.; *Biomater.* **2005,** *26,* 4139.
26. Fang, X.; and Reneker, D. H.; *J. Macromole. Sci. Part B: Phys.* **1997,** *36,* 169.
27. Taylor, G. I.; *Proc. Royal Soc. Ser. A.* **1969,** *313,* 453.
28. Kenawy, E. -R.; et al. *J. Contr. Rel.* **2002,** *81,* 57.
29. Fennessey, S. F.; and Farris, J. R.; *Polym.* **2004,** *45,* 4217.
30. Zussman, E.; Theron, A.; and Yarin, A. L.; *Appl. Phys. Lett.* **2003,** *82,* 973.
31. Deitzel, J. M.; Kleinmeyer, J.; Harris, D.; and Beck, T. N.; *Polym.* **2001,** *42,* 261.

32. Zhang, C. H.; Yuan, X.; Wu, L.; Han, Y.; and Sheng, J.; *Euro Polym. J.* **2005,** *41,* 423.
33. Spivak, A. F.; and Dzenis, Y. A.; *Appl. Phys. Lett.* **1998,** *73,* 3067.
34. Hohman, M. M.; Shin, M.; Rutledge, G.; and Brenner, M. P.; *Phys. Fluids.* **2001a,** *13,* 2201.
35. Hohman, M. M.; Shin, M.; Rutledge, G.; and Brenner, M. P.; *Phys. Fluids.* **2001b,** *13,* 2221.
36. Reneker, D. H.; Yarin, A. L.; Fong, H.; and Koombhongse, S.; *J. Appl. Phys.* **2000,** *87,* 4531.
37. Yarin, A. L.; Koombhongse, S.; and Reneker, D. H.; *J. Appl. Phys.* **2001,** *89,* 47.

CHAPTER 3

A NOTE ON BOND ENERGY CALCULATIONS IN NANOSTRUCTURED MATERIAL

A. K. HAGHI and G. E. ZAIKOV

CONTENTS

3.1   INTRODUCTION

Water plays an ambiguous role in hydrocarbon fuel of internal combustion engines. On the one hand, simple dilution of petroleum or diesel fuel with water can significantly deteriorate technological characteristics of the fuel. As soon as water drops enter cylinders, the following phenomenon occurs: in the compression stroke when both valves are closed, the piston bears against the water plug when moving upward. The pressure inside the cylinder increases rapidly. The engine continues the cycle of bringing the connecting rod to the upper position. In fact, the pistons in one or several cylinders stop at once, and the crankshaft that continues rotating takes enormous loads. It bends connecting rods, breaks piston pins, and often breaks down itself. On the other hand, optimal water content in hydrocarbon fuel is defined by the standard technological norm of such fuel mixture, which is prepared by a special technique. Moreover, water-containing fuel can have the potential energy, and nevertheless, the engines produce the same power as with additional amount of petroleum by the mass equaled to the mass of water added. Not only is there power gain but also benefit in fuel technological characteristics, such as fire safety, octane number, application temperature limits, possibility to use cheaper fuels, and so on. Such specificity of technological processes is ultimately defined by the mechanism of physicochemical transformations occurring on atom-molecular level [1, 2]. In this investigation, their possible evaluations are studied based on the concept of spatial-energy parameter (P-parameter).

3.2   FORMATION OF HIGH-ENERGY BONDS IN THE SYSTEM

Practical use of hydrogen-containing fuel is possible only if the following conditions are fulfilled:

1.   Introduction of complex additives into the fuel, with alcohols and so-called "hydrogen catalyst" content having the primary meaning.
2.   Such additives are mixed following the special technique—first, by separate fractions, and in the end, all the mixture is intensively stirred by a hydraulic cutting pump (hydraulic shears).

"Hydrogen catalyst" contributes to active dissociation of water molecules with the formation of hydrogen and oxygen, which burn in the engine chamber afterward.

However, it is not clear how during such a short combustion time of the given amount of mixture introduced into the chamber, water dissociation in this volume and burning of its products can take place. Moreover, as a result of water dissociation by the reaction $H_2O = H^+ + OH^-$, the direct oxygen release is not observed. Obviously, other important mechanisms of physicochemical transformation of energy are involved. For instance, it is known that as a result of biochemical reactions in the presence of certain ferments, the synthesis of ATP molecule can take place whose potential energy increases due to the formation of special high-energy bonds.

It is possible that similar processes take place during the formation of burning mixture of this type of fuel when nanoclusters in the form of fullerenes can be formed under certain technological conditions. First, this is aided by the introduction of alcohols into the fuel mixture that results in the formation of fullerene, for example, $C_{60}(OH)_{10}$. Therefore, the addition of alcohols (up to 20%) just corresponds to the ratio of molar masses of hydroxyl groups OH^- and carbon atoms. At the second stage of fuel preparation, high-energy bonds are formed in the systems $C_{60}(OH^-) - n(H_2O)$, first, due to the introduction of "hydrogen catalyst" into the mixture, and besides, when filtering water through the coal filter, which contributes to the extraction of nanostructured formations of carbon atoms into the mixture.

Similar to ATP hydrolysis, which is accompanied by the release of chemical bond energy, the breakage of high-energy bonds and heat energy release occur in hydrogen-containing fuel when it is burning in the engine chamber.

The physicochemical mechanism of the formation of energy-saturated bonds in this system is discussed in the following section.

3.3   EXPERIMENTAL

The value of the relative difference of P-parameters of interacting atom components—coefficient of structural interaction α was used as the major numerical characteristic of structural interactions in condensed media:

$$\alpha = \frac{P_1 - P_2}{(P_1 + P_2)/2} \cdot 100\% \qquad (3.1)$$

Applying the reliable experimental data, we obtain the nomogram of the dependence degree of structural interactions upon coefficient α—unified for a wide range of structures. This approach allows evaluating the degree and directedness of structural interactions of phase formation, isomorphism, and solubility processes in multiple systems, including molecular systems. In particular, the features of cluster formation in the system $CaSO_4-H_2O$ have been investigated.

To evaluate the directedness and degree of phase formation processes, the following equations are used:

1. Initial values of P-parameters:

$$\frac{1}{q^2 / r_A} + \frac{1}{W_i n_i} = \frac{1}{P} \qquad (3.2)$$

$$\frac{1}{P_0} = \frac{1}{q^2} + \frac{1}{(Wrn)_i} \qquad (3.3)$$

$$P_A = P_0 / r_i \qquad (3.4)$$

where W_i—electron orbital energy; r_i—orbital radius of i orbital; $q = Z^* / n^*$—; n_i—number of electrons of the given orbital, Z^* and n^*—effective nucleus charge and effective main quantum number. P_0 is the spatial-energy parameter and P_E the effective P-parameter.

The calculation results by equations for a number of elements are summarized in Table 3.1, from which it is seen that for hydrogen atom the values of P_E parameters substantially differ at the distance of orbital (r_i) and covalent (R) radii. The hybridization of valence orbitals of carbon atom is evaluated as the averaged value of P parameters of $2S^2$ and $2P^2$ orbitals.

2. Values of P_c parameter in binary and complex structures:

$$\frac{1}{P_c} = \frac{1}{N_1 P_1} + \frac{1}{N_2 P_2} + \ldots \ldots \qquad (3.5)$$

where N is the number of homogeneous atoms in each subsystem.

The results of such calculations for some systems are summarized in Table 3.2.

3. Bond energy (E) in binary and more complex systems:

$$\frac{1}{E} \approx \frac{1}{P_E} = \frac{1}{P_1(N/\kappa)_1} + \frac{1}{P_2(N/\kappa)_2} + \ldots\ldots\ldots \quad (3.6)$$

Here (as applicable to cluster systems) κ_1 and κ_2—number of subsystems forming the cluster system; N_1 and N_2—number of homogeneous clusters.

So, for $C_{60}(OH)_{10\ldots}$ $\kappa_1 = 60$, $\kappa_2 = 10$.

3.4 CALCULATIONS AND COMPARISONS

It was assumed that structural-stable water cluster (H_2O) can have the same static number of subsystems (κ) as the number of subsystems in the system interacting with it. For example, the water cluster $n(H_2O)_{10}$ is interacting with fullerene $[C_6OH]_{10}$.

Similarly, with cluster $[C_6OH]_{10}$, the formation of cluster $[(C_2H_5OH)_6-H_2O]_{10}$ is apparently possible, which corresponds to the system $(C_2H_5OH)_{60}-(H_2O)_{10}$. The interaction of water clusters was considered as the interaction of subsystems $(H_2O)_{60}-N(H_2O)_{60}$.

Based on such concepts, the bond energies in these systems are calculated by Eq. (3.6). The results are summarized in Table 3.3.

For comparison, the calculation data obtained by with quantum-chemical techniques are provided.

Both the techniques produce consistent values of bond energy (in eV). Besides, the methodology of P-parameter allows explaining why the energy of cluster bonds of water molecules with fullerene $C_{60}(OH)_{10}$ exceed two times the bond energy between the molecules of cluster water (Table 3.3).

In accordance with the nomogram, the phase formation of structures can take place only if the relative difference of their P-parameters (α) is under 25–30 per cent, and the most stable structures are formed when α <6 − 7 per cent.

In Table 3.4, different values of coefficient α in systems H–C, H–OH, and H–H$_2$O are provided, which are within 0.44 − 7.09(%).

But in the system H-C, for carbon and hydrogen atoms, the interactions at the distances of covalent radii have been taken into account, and for other systems—at the distance of orbital radius.

The interaction in system H-C at the distances of covalent radius plays a role of fermentative action, which results in the transition of dimensional characteristics in water molecules from the orbital radius to the covalent one and formation of system $C_{60}(OH)_{10} - N(H_2O)_{10}$ with bond energy between the main components two times greater than between the water molecules (high-energy bonds).

TABLE 3.1 P-parameters of atoms calculated via the electron bond energy

Atom	Valence electrons	W (eV)	r_i (Å)	q^2 (eV Å)	P_0(eV Å)	R (Å)	$P_E = P_0/R$ (eV)
H	1S^1	13.595	0.5295	14.394	4.7985	0.5295	9.0624
H	1S^1					0.28	17.137
C	2P^1	11.792	0.596	35.395	5.8680	0.77	7.6208
C	2P^1					0.69	8.5043
C	2P^2	11.792	0.596	35.395	10.061	0.77	13.066
C	2S^1	19.201	0.620	37.240	9.0209	0.77	11.715
C	2S^2				14.524	0.77	18.862
C	2S^2 + 2P^2				24.585	0.77	31.929
C	1/2(2S^2 + 2P^2)						15.964
O	2P^1	17.195	0.4135	71.383	4.663	0.66	9.7979
O	2P^2	17.195	0.4135	71.383	11.858	0.66	17.967
O	2P^2					0.59	20.048
O	2P^4	17.195	0.4135	71.383	20.338	0.66	30.815

TABLE 3.2 Structural P_c-parameters

Radicals, molecules	P_1 (eV)	P_2 (eV)	P_3 (eV)	P_4 (eV)	P_c (eV)	Orbitals of oxygen atom
OH	17.967	17.137			8.7712	$2P^2$
OH	9.7979	9.0624			4.7080	$2P^1$
H_2O	2×17.138	17.967			11.788	$2P^2$
H_2O	2×9.0624	17.967			9.0226	$2P^2$
C_2H_5OH	2×15.964	2×9.0624	9.7979	9.0624	3.7622	$2P^1$

TABLE 3.3 Calculation of bond energy—E (eV)

System	C_{60}	$(OH)_{10}$	$(H_2O)_{10}$		P_E (eV) (calculation)	E (eV)
	P_1/κ_1	P_2/κ_2	P_3/κ_3	n_3	By equation (3.6)	Quantum-mechanical
$C_{60}(OH)_{10}-N(H_2O)_{10}$	15.964/60	8.7712/10	11.788/10	1	0.174	0,176
				2	0.188	0,209
				3	0.193	0,218
				4	0.196	0,212
				5	0.197	0,204
$(H_2O)_{60} - N(H_2O)_{60}$	9.0226/60	9.0226/60		n_2		
				1	0.0768	0,0863
				2	0.1020	0,1032
				3	0.1128	0,1101
				4	0.1203	0,1110
				5	0.1274	0,115
$(C_2H_5OH)_{60}-(H_2O)_{10}$	3.7622/60	9.0226/10			0.0586	0.0607
$(C_2H_5OH)_{10}-(H_2O)_{60}$	3.7622/10	9.0226/60			0.1074	≈ 0.116

TABLE 3.4   Spatial-energy interactions in the system H-R, where R = C, (OH), H$_2$O

System	P$_1$(eV)	P$_2$(eV)	$\alpha = \dfrac{\Delta P}{\langle P \rangle} 100\%$	Spatial bond type
H–C	17.137	15.964	7.09	Covalent
H–OH	9.0624	8.7712	3.27	Orbital
H–H$_2$O	9.0624	9.0226	0.44	Orbital

Thus, broad capabilities of water clusters to change their spatial-energy characteristics apparently explain all the diversity of structural properties of water in its different modifications, including the formation of high-energy bonds in water-containing fuel for internal combustion engines.

3.5   SUMMARY

The consistent calculations of bond energy in cluster water nanostructures have been performed following the P-parameter methodology and quantum-mechanical methods.

The formation of high-energy bonds in the process of hydrocarbon hydrogen containing fuel preparation has been explained.

1. Results of bond energy calculations in water cluster nanostructures following the P-parameter methodology agree with quantum-mechanical methods.
2. Changes that can take place in spatial-energy characteristics of water clusters explain the formation of high-energy bonds in the process of hydrocarbon fuel preparation.
3. Breaking of these bonds with the release of additional amount of heat energy occurs in the combustion chamber.

KEYWORDS

- **Bond energy**
- **Calculations**
- **Nanostructured materials**

REFERENCES

1. Tewarson, A.; Su, P.; and Yee, G. G.; In: The Proceedings of the Interscience Communications Conference—Hazards of Combustion Products: Toxicity, Opacity, Corrosivity and Heat Release. Eds. Babrauskas, V.; Gann, R.; and Grayson, S.; Greenwich, UK; **2008,** 225 p.
2. Warneck, P.; In: Chemistry of Multiphase Atmospheric Systems. Ed. Jaeschle, W.; NATO ASI Series. Berlin, Germany: Springer-Verlag; **1986,** *6G,* 473 p.

CHAPTER 4

A COMPREHENSIVE REVIEW ON COMPUTATIONAL METHODS FOR NANOPOROUSPOLYMER/ CARBON NANOTUBE MEMBRANE SEPARATION

AREZO AFZALI and SHIMA MAGHSOODLOU

CONTENTS

4.1   MEMBRANES FILTRATION

Membrane filtration is a mechanical filtration technique which uses an absolute barrier to the passage of particulate material as any technology currently available in water treatment. The term "membrane" covers a wide range of processes, including those used for gas/gas, gas/liquid, liquid/liquid, gas/solid, and liquid/solid separations. Membrane production is a large-scale operation. There are two basic types of filters: depth filters and membrane filters.

Depth filters have a significant physical depth and the particles to be maintained are captured throughout the depth of the filter. Depth filters often have a flexuous three-dimensional structure, with multiple channels and heavy branching so that there is a large pathway through which the liquid must flow and by which the filter can retain particles. Depth filters have the advantages of low cost, high throughput, large particle retention capacity, and the ability to retain a variety of particle sizes. However, they can endure from entrainment of the filter medium, uncertainty regarding effective pore size, some ambiguity regarding the overall integrity of the filter, and the risk of particles being mobilized when the pressure differential across the filter is large.

The second type of filter is the membrane filter, in which depth is not considered momentous. The membrane filter uses a relatively thin material with a well-defined maximum pore size and the particle retaining effect takes place almost entirely at the surface. Membranes offer the advantage of having well-defined effective pore sizes, can be integrity tested more easily than depth filters, and can achieve more filtration of much smaller particles. They tend to be more expensive than depth filters and usually cannot achieve the throughput of a depth filter. Filtration technology has developed a well-defined terminology that has been well addressed by commercial suppliers.

The term membrane has been defined in a number of ways. The most appealing definitions to us are the following:

"A selective separation barrier for one or several components in solution or suspension" [19]. "A thin layer of material that is capable of separating materials as a function of their physical and chemical properties when a driving force is applied across the membrane."

Membranes are important materials which form part of our daily lives. Their long history and use in biological systems has been extensively studied throughout the scientific field. Membranes have proven themselves as promising separation candidates due to advantages offered by their high stability, efficiency, low energy requirement and ease of operation. Membranes with good thermal and mechanical stability combined with good solvent resistance are important for industrial processes [1].

The concept of membrane processes is relatively simple but nevertheless often unknown. Membranes might be described as conventional filters but with much finer mesh or much smaller pores to enable the separation of tiny particles, even molecules. In general, one can divide membranes into two groups: porous and nonporous. The former group is similar to classical filtration with pressure as the driving force; the separation of a mixture is achieved by the rejection of at least one component by the membrane and passing of the other components through the membrane (see Figure4.1). However, it is important to note that nonporous membranes do not operate on a size exclusion mechanism.

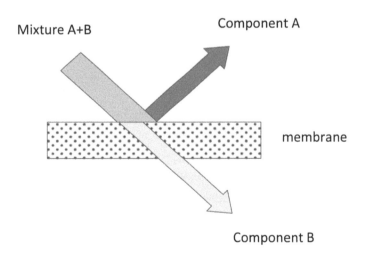

FIGURE 4.1   Basic principle of porous membrane processes.

Membrane-separation processes can be used for a wide range of applications and can often offer significant advantages over conventional separation such as distillation and adsorption since the separation is based on a physical mechanism. Compared to conventional processes, therefore, no chemical, biological, or thermal change of the component is involved for most membrane processes. Hence membrane separation is particularly attractive to the processing of food, beverage, and bio products where the processed products can be sensitive to temperature (vs. distillation) and solvents (vs. extraction).

Synthetic membranes show a large variety in their structural forms. The material used in their production determines their function and their driving forces. Typically the driving force is pressure across the membrane barrier (see Table 4.1) [2–4]]. Formation of a pressure gradient cross the membrane allows separation in a bolter-like manner. Some other forms of separation that exist include charge effects and solution diffusion. In this separation, the smaller particles are allowed to pass through as permeates whereas the larger molecules (macromolecules) are retained. There tention or permeation of these species is ordained by the pore architecture as well as pore sizes of the membrane employed. Therefore based on the pore sizes, these pressure-driven membranes can be divided into reverse osmosis (RO), nanofiltration (NF), ultrafiltration (UF), and microfiltration (MF), are already applied on an industrial scale to food and bioproduct processing [5–7].

TABLE 4.1 Drivingforces and their membrane processes

Driving force	Membrane process
Pressure difference	Microfiltration, Ultrafiltration, Nanofiltration, Reverseosmosis
Chemical potential difference	Pervaporation, Pertraction, Dialysis, Gas separation, Vapor permeation, Liquid membranes
Electrical potential difference	Electrodialysis, Membrane electrophoresis, Membrane electrolysis
Temperature difference	Membrane distillation

Microfiltration (MF) membranes

MF membranes have the largest pore sizes and thus use less pressure. They involve removing chemical and biological species with diameters ranging between 100 to10,000 nm and components smaller than this, pass through as permeates. MF is primarily used to separate particles and bacteria from other smaller solutes [4].

A. Ultrafiltration (UF) membranes

UF membranes operate within the parameters of the micro- and nano-filtration membranes. Therefore UF membranes have smaller pores as compared to MF membranes. They involve retaining macromolecules and colloids from solution which range between 2–100 nm and operating pressures between 1 and 10 bar (e.g., large organic molecules and proteins). UF is used to separate colloids such as proteins from small molecules such as sugars and salts [4].

B. Nanofiltration (NF) membranes

NF membranes are distinguished by their pore sizes of between 0.5–2 nm and operating pressures between 5 and 40 bar. They are mainly used for the removal of small organic molecules and di- and multivalent ions. Additionally, NF membranes have surface charges that make them suitable for retaining ionic pollutants from solution. NF is used to achieve separation between sugars, other organic molecules, and multivalent salts on the one hand from monovalent salts and water on the other. Nanofiltration, however, does not remove dissolved compounds [4].

C. Reverse osmosis (RO) membranes

RO membranes are dense semi-permeable membranes mainly used for desalination of sea water [38]. Contrary to MF and UF membranes, RO membranes have no distinct pores. As a result, high pressures are applied to increase the permeability of the membranes [4]. The properties of the various types of membranes are summarized in Table4.2.

TABLE 4.2 Summary of properties of pressure-driven membranes [4]

	MF	UF	NF	RO
Permeability (L/h. m².bar)	1,000	10–1,000	1.5–30	0.05–1.5
Pressure (bar)	0.1–2	0.1–5	3–20	5–1,120
Pore size (nm)	100–10,000	2–100	0.5–2	0.5
Separation Mechanism	sieving	sieving	Sieving, charge Effects	Solution diffusion
Applications	Removal of bacteria	Removal of bacteria, fungi, virses	Removal of multivalent ions	Desalinatiob

The NF membrane is a type of pressure-driven membrane with properties in between RO and UF membranes. NF offers several advantages such as low operation pressure, high flux, high retention of multivalent anion salts and an organic molecular above 300, relatively low investment and low operation and maintenance costs. Because of these advantages, the applications of NF worldwide have increased [8]. In recent times, research in the application of nanofiltration techniques has been extended from separation of aqueous solutions to separation of organic solvents to homogeneous catalysis, separation of ionic liquids, food processing, etc [9].

Figure 4.2 presents a classification on the applicability of different membrane-separation processes based on particle or molecular sizes. RO process is often used for desalination and pure water production, but it is the UF and MF that are widely used in food and bioprocessing.

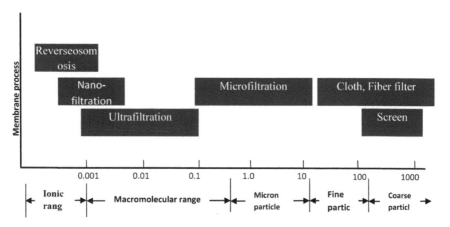

FIGURE 4.2   The applicability ranges of different separation processes based on sizes.

While MF membranes target on the microorganism removal, and hence are given the absolute rating, namely, the diameter of the largest pore on the membrane surface (Figure 4.3), UF/NF membranes are characterized by the nominal rating due to their early applications of purifying biological solutions. The nominal rating is defined as the molecular weight cut-off (MWCO) that is the smallest molecular weight of species, of which the membrane has more than 90 per cent rejection (see later for definitions). The separation mechanism in MF/UF/NF is mainly the size exclusion, which is indicated in the nominal ratings of the membranes. The other separation mechanism includes the electrostatic interactions between solutes and membranes, which depends on the surface and physiochemical properties of solutes and membranes[5]. Also, The principal types of membrane are shown schematically in Figure4.4 and are described briefly below.

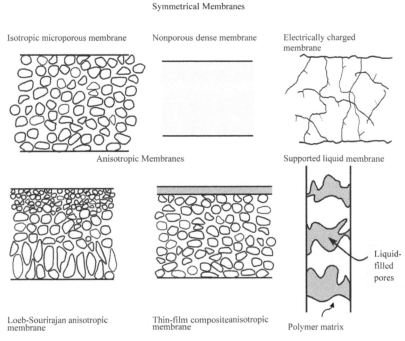

FIGURE 4.3   Schematic diagrams of the principal types of membranes.

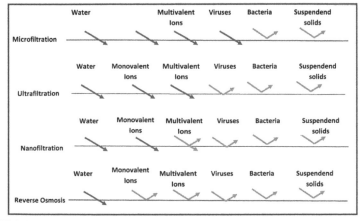

FIGURE 4.4   Membrane process characteristics.

4.2   THE RELATIONSHIP BETWEEN NANOTECHNOLOGY AND FILTRATION

Nowadays, nanomaterials have become the most interested topic of materials research and development due to their unique structural properties (unique chemical, biological, and physical properties as compared to larger particles of the same material) that cover their efficient uses in various fields, such as ion exchange and separation, catalysis, biomolecular isolation and purification as well as in chemical sensing [10]. However, the understanding of the potential risks (health and environmental effects) posed by nanomaterials hasn't increased as rapidly as research has regarding possible applications.

One of the ways to enhance their functional properties is to increase their specific surface area by the creation of a large number of nanostructured elements or by the synthesis of a highly porous material.

Classically, porous matter is seen as material containing three-dimensional voids, representing translational repetition, while no regularity is necessary for a material to be termed "porous". In general, the pores can be classified into two types: open pores which connect to the surface of the material, and closed pores which are isolated from the outside. If the material exhibits mainly open pores, which can be easily transpired, then one can consider its use in functional applications such as adsorption, catalysis and sensing. In turn, the closed pores can be used in sonic and thermal insulation, or lightweight structural applications. The use of porous materials offers also new opportunities in such areas as coverage chemistry, guest–host synthesis and molecular manipulations and reactions for manufacture of nanoparticles, nanowires and other quantum nanostructures. The International Union of Pure and Applied Chemistry (IUPAC) defines porosity scales as follows (Figure 4.5):

- Microporous materials 0–2-nm pores
- Mesoporous materials 2–50-nm pores
  Macroporous materials >50-nm pores

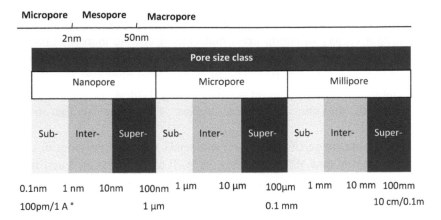

FIGURE 4.5 New pore size classification as compared with the current IUPAC nomenclature.

This definition, it should be noted, is somewhat in conflict with the definition of nanoscale objects, which typically have large relative porosities (>0.4), and pore diameters between 1 and 100 nm. In order to classify porous materials according to the size of their pores the sorption analysis is one of the tools often used. This tool is based on the fact that pores of different sizes lead to totally different characteristics in sorption isotherms. The correlation between the vapor pressure and the pore size can be written as the Kelvin equation:

$$r_p\left(\frac{p}{p_0}\right) = \frac{2\gamma V_L}{RT\ln\left(\frac{p}{p_0}\right)} + t\left(\frac{p}{p_0}\right) \tag{4.1}$$

Therefore, the isotherms of microporous materials show a steep increase at very low pressures (relative pressures near zero) and reach a plateau quickly. Mesoporous materials are characterized by a so-called capillary doping step and a hysteresis (a discrepancy between adsorption and desorption). Macroporous materials show a single or multiple adsorption steps near the pressure of the standard bulk condensed state (relative pressure approaches one) [10].

Nanoporous materials exuberate in nature, both in biological systems and in natural minerals. Some nanoporous materials have been used industrially for a longtime. Recent progress in characterization and manipulation on the nanoscale has led to noticeable progression in understanding and making a variety of nanoporous materials: from the merely opportunistic to directed design. This is most strikingly the case in the creation of a wide variety of membranes where control over pore size is increasing dramatically, often to atomic levels of perfection, as is the ability to modify physical and chemical characteristics of the materials that make up the pores [11].

The available range of membrane materials includes polymeric, carbon, silica, zeolite and other ceramics, as well as composites. Each type of membrane can have a different porous structure, as illustrated in Figure4.6. Membranes can be thought of as having a fixed (immovable)network of pores in which the molecule travels, with the exception of most polymeric membranes [12–13]. Polymeric membranes are composed of an amorphous mix of polymer chains whose interactions involve mostly Van der Waals forces. However, some polymers manifest a behavior that is consistent with the idea of existence of opened pores within their matrix. This is especially true for high free volume, high permeability polymers, as has been proved by computer modeling, low activation energy of diffusion, negative activation energy of permeation, solubility controlled permeation [14–15]. Although polymeric membranes have often been viewed as nonporous, in the modeling framework discussed here it is convenient to consider them nonetheless as porous. Glassy polymers have pores that can be considered as 'frozen' over short times scales, while rubbery polymers have dynamic fluctuating pores (or more correctly free volume elements) that move, shrink, expand and disappear [16].

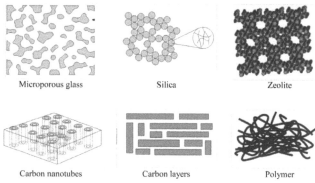

Microporous glass   Silica   Zeolite

Carbon nanotubes   Carbon layers   Polymer

FIGURE 4.6   Porous structure within various types of membranes.

Three nanotechnologies that are often used in the filtering processes and show great potential for applications in remediation are:

1. Nanofiltration (and its sibling technologies: reverse osmosis, ultrafiltration, and microfiltration), is a fully developed, commercially-available membrane technology with a large number of vendors. Nanofiltration relies on the ability of membranes to discriminate between the physical size of particles or species in a mixture or solution and is primarily used for water pre-treatment, treatment, and purification). There are almost 600 companies in worldwide which offering membrane systems.

2. Electro spinning is a process utilized by the nanofiltration process, in which fibers are stretched and elongated down to a diameter of about 10 nm. The modified nanofibers that are produced are particularly useful in the filtration process as an ultra-concentrated filter with a very large surface area. Studies have found that electro spun nanofibers can capture metallic ions and are continually effective through re-filtration.

3. Surface modified membrane is a term used for membranes with altered makeup and configuration, though the basic properties of their underlying materials remain intact.

4.3 TYPES OF MEMBRANES

As it mentioned, membranes have achieved a momentous place in chemical technology and are used in a broad range of applications. The key property that is exploited is the ability of a membrane to control the permeation rate of a chemical species through the membrane. In essence, a membrane is nothing more than a discrete, thin interface that moderates the permeation of chemical species in contact with it. This interface may be molecularly homogeneous, that is completely uniform in composition and structure or it may be chemically or physically heterogeneous for example, containing holes or pores of finite dimensions or consisting of some form of layered structure. A normal filter meets this definition of a membrane, but, generally, the term filter is usually limited to structures that separate particulate suspensions larger than 1–10 μm [17].

The preparation of synthetic membranes is however more recent invention which has received a great audience due to its applications [18].

Membrane technology like most other methods has undergone a developmental stage, which has validated the technique as a cost-effective treatment option for water. The level of performance of the membrane technologies is still developing and it is stimulated by the use of additives to improve the mechanical and thermal properties, as well as the permeability, selectivity, rejection, and fouling of the membranes [19]. Membranes can be fabricated to possess different morphologies. However, most membranes that have found practical use are mainly of asymmetric structure. Separation in membrane processes takes place as a result of differences in the transport rates of different species through the membrane structure, which is usually polymeric or ceramic [20].

The versatility of membrane filtration has allowed their use in many processes where their properties are suitable in the feed stream. Although membrane separation does not provide the ultimate solution to water treatment, it can be economically connected to conventional treatment technologies by modifying and improving certain properties [21].

The performance of any polymeric membrane in a given process is highly dependent on both the chemical structure of the matrix and the physical arrangement of the membrane [22]. Moreover, the structural impeccability of a membrane is very important since it determines its permeation and selectivity efficiency. As such, polymer membranes should be seen as much more than just sieving filters, but as intrinsic complex structures which can either be homogenous (isotropic) or heterogeneous (anisotropic), porous or dense, liquid or solid, organic or inorganic [22–23].

4.3.1   ISOTROPIC MEMBRANES

Isotropic membranes are typically homogeneous/uniform in composition and structure. They are divided into three subgroups, namely: microporous, dense and electrically charged membranes[20]. Isotropic microporous membranes have evenly distributed pores (Figure4.7(a)) [27]. Their pore diameters range between 0.01 and 10 μm and operate by the sieving mechanism. The microporous membranes are mainly prepared by the phase inversion method albeit other methods can be used. Conversely, isotropic dense membranes do not have pores and as a result they tend to be thicker than the microporous membranes (Figure 4.7(b)). Solutes are carried through the membrane by diffusion under a pressure, concentration or electrical potential gradient. Electrically charged membranes can either be porous or non-porous. However in most cases they are finely microporous with pore walls containing charged ions (Figure 4.7(c)) [20, 28].

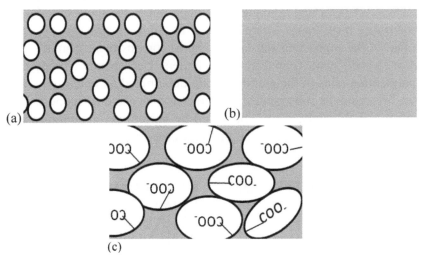

FIGURE 4.7 Schematic diagram so isotropic membranes:(a) microporous; (b) dense; and (c) electrically charged membranes.

4.3.2 ANISOTROPIC MEMBRANES

Anisotropic membranes are often referred to as Loeb-Sourirajan, based on the scientists who first synthesized them [24–25]. They are the most widely used membranes in industries. The transport rate of a species through a membrane is inversely proportional to the membrane thickness. The membrane should be as thin as possible due to high transport rates are eligible in membrane-separation processes for economic reasons. Contractual film fabrication technology limits manufacture of mechanically strong, defect-free films to thicknesses of about 20 μm. The development of novel membrane fabrication techniques to produce anisotropic membrane structures is one of the major breakthroughs of membrane technology. Anisotropic membranes consist of an extremely thin surface layer supported on a much thicker, porous substructure. The surface layer and its substructure may be formed in a single operation or separately [17]. They are represented by non-uniform structures which consist of a thin active skin layer and a highly porous support layer. The active layer enjoins the efficiency of the membrane, whereas the porous support layer influences the mechanical stability of the membrane. An isotropic membrane (Figure 4.8) can be classified into two groups, namely: (i) integrally skinned membranes where the active layer is formed from the same substance as the supporting layer, (ii) composite membranes where the polymer of the active layer differs from that of the supporting sub-layer [25]. In composite membranes, the layers are usually made from different polymers. The separation properties and permeation rates of the membrane are determined particularly by the surface layer and the substructure functions as a mechanical support. The advantages of the higher fluxes provided by anisotropic membranes are so great that almost all commercial processes use such membranes [17].

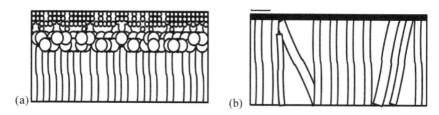

FIGURE 4.8 Schematic diagrams of an isotropic membranes: (a) Loeb-Sourirajan and (b) thin-film composite membranes.

4.3.3 POROUS MEMBRANE

In Knudsen diffusion (Figure 4.9(a)), the pore size forces the penetrant molecules to collide more frequently with the pore wall than with other incisive species [26]. Except for some special applications as membrane reactors, Knudsen-selective membranes are not commercially attractive because of their low selectivity [27]. In surface diffusion mechanism (Figure 4.9(b)), the pervasive molecules adsorb on the surface of the pores to move from one site to another of lower concentration. Capillary condensation (Figure 4.9(c)) impresses the rate of diffusion across the membrane. It occurs when the pore size and the interactions of the penetrant with the pore walls induce penetrant condensation in the pore [28]. Molecular-sieve membranes in Figure 4.9(d) have gotten more attention because of their higher productivities and selectivity than solution-diffusion membranes.

Molecular sieving membranes are means to polymeric membranes. They have ultramicroporous (<7Å) with sufficiently small pores to barricade some molecules, while allowing others to pass through. Although they have several advantages such as permeation performance, chemical and thermal stability, they are still difficult to process because of some properties like fragile. Also they are expensive to fabricate.

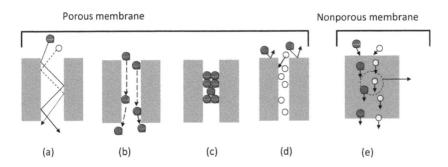

FIGURE 4.9 Schematic representation of membrane-based gas separations. (a) Knudsen-flow separation, (b) surface-diffusion, (c) capillary condensation, (d) molecular-sieving separation, and (e) solution-diffusion mechanism.

4.3.4 NONPOROUS (DENSE) MEMBRANE

Nonporous, dense membranes consist of a dense film through which permeants are transported by diffusion under the driving force of a pressure, concentration, or electrical potential gradient. The separation of various components of a mixture is related directly to their relative transport rate within the membrane, which is determined by their diffusivity and solubility in the membrane material. Thus, nonporous, dense membranes can separate permeants of similar size if the permeant concentrations in the membrane material differ substantially. Reverse osmosis membranes use dense membranes to perform the separation. Usually these membranes have an anisotropic structure to improve the flux [17].

The mechanism of separation by non-porous membranes is different from that by porous membranes. The transport through nonporous polymeric membranes is usually described by a solution-diffusion mechanism (Figure 4.9(e)). The most current commercial polymeric membranes operate according to the solution-diffusion mechanism. The solution-diffusion mechanism has three steps: (1) the absorption or adsorption at the upstream boundary, (2) activated diffusion through the membrane, and (3) desorption or evaporation on the other side. This solution-diffusion mechanism is driven by a difference in the thermodynamic activities existing at the upstream and downstream faces of the membrane as well as the intermolecular forces acting between the permeating molecules and those making up the membrane material.

The concentration gradient causes the diffusion in the direction of decreasing activity. Differences in the permeability in dense membranes are caused not only by diffusivity differences of the various species but also by differences in the physicochemical interactions of the species within the polymer. The solution-diffusion model assumes that the pressure within a membrane is uniform and that the chemical potential gradient across the membrane is expressed only as a concentration gradient. This mechanism controls permeation in polymeric membranes for separations.

4.4   CARBON NANOTUBES-POLYMER MEMBRANE

Iijima discovered carbon nanotubes (CNTs) in 1991 and it was really a revolution in nanoscience because of their distinguished properties. CNTs have the unique electrical properties and extremely high thermal conductivity [29–30] and high elastic modulus (>1TPa), large elastic strain—up to 5per cent, and large breaking strain—up to 20per cent. Their excellent mechanical properties could lead to many applications[31]. For example, with their amazing strength and stiffness, plus the advantage of lightness, perspective future applications of CNTs are in aerospace engineering and virtual biodevices [32].

CNTs have been studied worldwide by scientists and engineers since their discovery, but a robust, theoretically precise and efficient prediction of the mechanical properties of CNTs has not yet been found. The problem is, when the size of an object is small to nanoscale, their many physical properties cannot be modeled and analyzed by using constitutive laws from traditional continuum theories, since the complex atomistic processes affect the results of their macroscopic behavior. Atomistic simulations can give more precise modeled results of the underlying physical properties. Due to atomistic simulations of a whole CNT are computationally infeasible at present, a new atomistic and continuum mixing modeling method is needed to solve the problem, which requires crossing the length and time scales. The research here is to develop a proper technique of spanning multi-scales from atomic to macroscopic space, in which the constitutive laws are derived from empirical atomistic potentials which deal with individual interactions between single atoms at the micro-level, whereas Cosserat continuum theories are adopted for a shell model through the application of the Cauchy-Born rule to give the properties which represent the averaged behavior of large volumes of atoms at the macro-level [33–34]. Since experiments of CNTs are relatively expensive at present, and often unexpected manual errors could be involved, it will be very helpful to have a mature theoretical method for the study of mechanical properties of CNTs. Thus, if this research is successful, it could also be a reference for the research of all sorts of research at the nanoscale, and the results can be of interest to aerospace, biomedical engineering [35].

Subsequent investigations have shown that CNTs integrate amazing rigid and tough properties, such as exceptionally high elastic properties, large elastic strain, and fracture strain sustaining capability, which seem inconsistent and impossible in the previous materials. CNTs are the strongest fibers known. The Young's Modulus of SWNT is around 1TPa, which is 5 times greater than steel (200 GPa) while the density is only 1.2~1.4 g/cm^3. This means that materials made of nanotubes are lighter and more durable.

Beside their well-known extra-high mechanical properties, single-walled carbon nanotubes (SWNTs) offer either metallic or semiconductor characteristics based on the chiral structure of fullerene. They possess superior thermal and electrical properties so SWNTs are regarded as the most promising reinforcement material for the next generation of high performance structural and multifunctional composites, and evoke great interest in polymer-based composites research. The SWNTs/polymer composites are theoretically predicted to have both exceptional mechanical and functional properties, which carbon fibers cannot offer [36].

4.4.1 CARBON NANOTUBES

Nanotubular materials are important "building blocks" of nanotechnology, in particular, the synthesis and applications of CNTs [37–39]. One application area has been the use of carbon nanotubes for molecular separations, owing to some of their unique properties. One such important property, extremely fast mass transport of molecules within carbon nanotubes associated with their low friction inner nanotube surfaces, has been demonstrated via computational and experimental studies [40–41]. Furthermore, the behavior of adsorbate molecules in nano-confinement is fundamentally different than in the bulk phase, which could lead to the design of new sorbents [42].

Finally, their one-dimensional geometry could allow for alignment in desirable orientations for given separation devices to optimize the mass transport. Despite possessing such attractive properties, several intrinsic limitations of carbon nanotubes inhibit their application in large-scale separation processes: the high cost of CNT synthesis and membrane formation (by micro fabrication processes), as well as their lack of surface functionality, which significantly limits their molecular selectivity [43]. Although outer-surface modification of carbon nanotubes has been developed for nearly two decades, interior modification via covalent chemistry is still challenging due to the low reactivity of the inner-surface. Specifically, forming covalent bonds at inner walls of carbon nanotubes requires a transformation from sp^2 to sp^3 hybridization. The formation of sp^3 carbon is energetically unfavorable for concave surfaces [44].

Membrane is a potentially effective way to apply nanotubular materials in industrial-scale molecular transport and separation processes. Polymeric membranes are already prominent for separations applications due to their low fabrication and operation costs. However, the main challenge for utilizing polymer membranes for future high-performance separations is to overcome the tradeoff between permeability and selectivity. A combination of the potentially high throughput and selectivity of nanotube materials with the process ability and mechanical strength of polymers may allow for the fabrication of scalable, high-performance membranes [45–46].

4.4.2 STRUCTURE OF CARBON NANOTUBES

Two types of nanotubes exist in nature: multi-walled carbon nanotubes (MWNTs), which were discovered by Iijima in 1991 [39] and SWNTs, which were discovered by Bethune et al. in 1993 [47–48].

Single-wall nanotube has only one single layer with diameters in the range of 0.6–1nm and densities of 1.33–1.40 g/cm^3[49]MWNTs are simply composed of concentric SWNTs with an inner diameter is from 1.5 to 15 nm and the outer diameter is from 2.5 to 30 nm [50]. SWNTs have better defined shapes of cylinder than MWNT, thus MWNTs have more possibilities of structure defects and their nanostructure is less stable. Their specific mechanical and electronic properties make them useful for future high strength/modulus materials and nanodevices. They exhibit low density, large elastic limit without breaking (of up to 20–30 per cent strain before failure), exceptional elastic stiffness, greater than 1,000 GP and their extreme strength which is more than twenty times higher than a high-strength steel alloy. Besides, they also possess superior thermal and elastic properties: thermal stability up to 2,800°C in vacuum and up to 750°C in air, thermal conductivity about twice as high as diamond, electric current carrying capacity 1,000 times higher than copper wire [51]. The properties of CNTs strongly depend on the size and the chirality and dramatically change when SWCNTs or MWCNTs are considered [52].

CNTs are formed from pure carbon bonds. Pure carbons only have two covalent bonds: sp^2 and sp^3. The former constitutes graphite and the latter constitutes diamond. The sp^2 hybridization, composed of one's orbital and two p orbitals, is a strong bond within a plane but weak between planes. When more bonds come together, they form six-fold structures, like honeycomb pattern, which is a plane structure, the same structure as graphite[53].

Graphite is stacked layer by layer so it is only stable for one single sheet. Wrapping these layers into cylinders and joining the edges, a tube of graphite is formed, called nanotube[54].

Atomic structure of nanotubes can be described in terms of tube chirality, or helicity, which is defined by the chiral vector, and the chiral angle, θ. Figure 4.10shows visualized cutting a graphite sheet along the dotted lines and rolling the tube so that the tip of the chiral vector touches its tail. The chiral vector, often known as the roll-up vector, can be described by the following equation [55]:

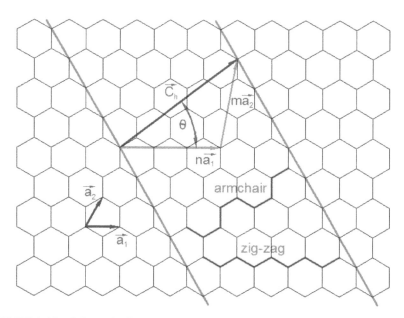

FIGURE 4.10 Schematic diagram showing how graphite sheet is 'rolled' to form CNT.

$$C_h = na_1 + ma_2 \tag{4.2}$$

As shown in Figure 4.10, the integers (n, m) are the number of steps along the carbon bonds of the hexagonal lattice. Chiral angle determines the amount of "twist" in the tube. Two limiting cases exist where the chiral angle is at 0° and 30°. These limiting cases are referred to as zig-zag (0°) and armchair (30°), based on the geometry of the carbon bonds around the circumference of the nanotube. The difference in armchair and zig-zag nanotube structures is shown in Figure 4.11 In terms of the roll-up vector, the zig-zag nanotube is (n, 0) and the armchair nanotube is (n, n). The roll-up vector of the nanotube also defines the nanotube diameter since the inter-atomic spacing of the carbon atoms is known [36].

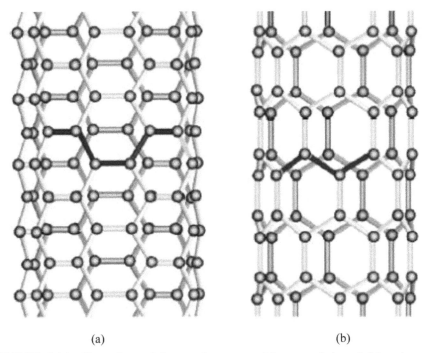

(a)   (b)

FIGURE 4.11 Illustrations of the atomic structure (a) an armchair and (b) a zig-zag nanotube.

Chiral vector C_h is a vector that maps an atom of one end of the tube to the other. C_h can be an integer multiple a_1 of a_2, which are two basis vectors of the graphite cell. Then we have $C_h = a_1 + a_2$, with integer n and m, and the constructed CNT is called a (n,m) CNT, as shown in Figure 4.12. It can be proved that for armchair CNTs n=m, and for zigzag CNTs m=0. In Figure 4.12, the structure is designed to be a (4,0) zigzag SWCNT.

MWCNT can be considered as the structure of a bundle of concentric SWCNTs with different diameters. The length and diameter of MWCNTs are different from those of SWCNTs, which means, their properties differ significantly. MWCNTs can be modeled as a collection of SWCNTs, provided the interlayer interactions are modeled by Van der Waals forces in the simulation. A SWCNT can be modeled as a hollow cylinder by rolling a graphite sheet as presented in Figure 4.13.

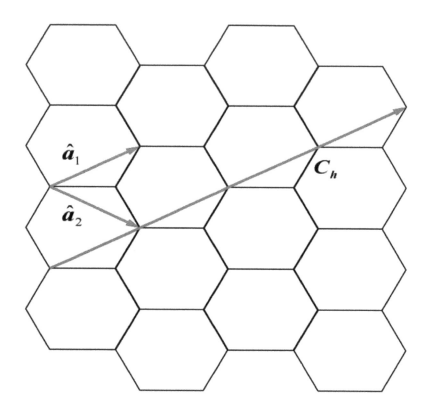

FIGURE 4.12   Basis vectors and chiral vector.

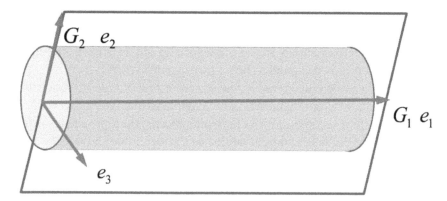

FIGURE 4.13   Illustration of a graphite sheet rolling to SWCNT.

If a planar graphite sheet is considered to be an undeformed configuration, and the SWCNT is defined as the current configuration, then the relationship between the SWCNT and the graphite sheet can be shown to be:

$$e_1 = G_1, e_2 = R\sin\frac{G_2}{R}, e_3 = R\cos\frac{G_2}{R} - R \qquad (4.3)$$

The relationship between the integer's n, m and the radius of SWCNT is given by:

$$R = a\sqrt{m^2 + mn + n^2} / 2\pi \qquad (4.4)$$

where $a = \sqrt{3}a_0$, and a_0 is the length of a non-stretched C-C bond which is 0.142nm [56].

As a graphite sheet can be 'rolled' into a SWCNT, we can 'unroll' the SWCNT to a plane graphite sheet. Since a SWCNT can be considered as a rectangular strip of hexagonal graphite monolayer rolling up to a cylindrical tube, the general idea is that it can be modeled as a cylindrical shell, a cylinder surface, or it can pull-back to be modeled as a plane sheet deforming into curved surface in three-dimensional space. A MWCNT can be modeled as a combination of a series of concentric SWCNTs with inter-layer inter-atomic reactions. Provided the continuum shell theory captures the deformation at the macro-level, the inner micro-structure can be described by finding the appropriate form of the potential function which is related to the position of the atoms at the atomistic level. Therefore, the SWCNT can be considered as a generalized continuum with microstructure [35].

4.4.3 CNT COMPOSITES

CNT composite materials cause significant development in nanoscience and nanotechnology. Their remarkable properties offer the potential for fabricating composites with substantially enhanced physical properties including conductivity, strength, elasticity, and toughness. Effective utilization of CNT in composite applications is dependent on the homogeneous distribution of CNTs throughout the matrix. Polymer-based nanocomposites are being developed for electronics applications such as thin-film capacitors in integrated circuits and solid polymer electrolytes for batteries.

Research is being conducted throughout the world targeting the application of carbon nanotubes as materials for use in transistors, fuel cells, big TV screens, ultra-sensitive sensors, high-resolution Atomic Force Microscopy (AFM) probes, super-capacitor, transparent conducting film, drug carrier, catalysts, and composite material. Nowadays, there are more reports on the fluid transport through porous CNTs/polymer membrane.

4.4.4 STRUCTURAL DEVELOPMENT IN POLYMER/CNT FIBERS

The inherent properties of CNT assume that the structure is well preserved (large-aspect-ratio and without defects). The first step toward effective reinforcement of polymers using nano-fillers is to achieve a uniform dispersion of the fillers within the hosting matrix, and this is also related to the as-synthesized nano-carbon structure. Secondly, effective interfacial interaction and stress transfer between CNT and polymer is essential for improved mechanical properties of the fiber composite. Finally, similar to polymer molecules, the excellent intrinsic mechanical properties of CNT can be fully exploited only if an ideal uniaxial orientation is achieved. Therefore, during the fabrication of polymer/CNT fibers, four key areas need to be addressed and understood in order to successfully control the micro-structural development in these composites. These are: (i) CNT pristine structure, (ii) CNT dispersion, (iii) polymer-CNT interfacial interaction and (iv) orientation of the filler and matrix molecules (Figure 4.14). Figure 4.14 Four major factors affecting the micro-structural development in polymer/CNT composite fiber during processing [57].

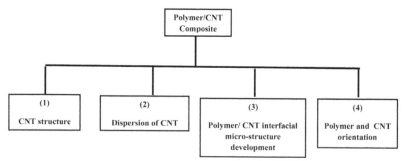

FIGURE 4.14   Four major factors affecting the micro-structural development in polymer/CNT composite fiber during processing.

Achieving homogenous dispersion of CNTs in the polymer matrix through strong interfacial interactions is crucial to the successful development of CNT/polymer nanocomposite [58]. As a result, various chemical or physical modifications can be applied to CNTs to improve its dispersion and compatibility with polymer matrix. Among these approaches acid treatment is considered most convenient, in which hydroxyl and carboxyl groups generated would concentrate on the ends of the CNT and at defect sites, making them more reactive and thus better dispersed [59–60].

The incorporation of functionalized CNTs into composite membranes are mostly carried out on flat sheet membranes [61–62]. For considering the potential influences of CNTs on the physicochemical properties of dope solution [63]and change of membrane formation route originated from various additives [64], it is necessary to study the effects of CNTs on the morphology and performance.

4.4.5 GENERAL FABRICATION PROCEDURES FOR POLYMER/ CNT FIBERS

In general, when discussing polymer/CNT composites, two major classes come to mind. First, the CNT nano-fillers are dispersed within a polymer at a specified concentration, and the entire mixture is fabricated into a composite. Secondly, as grown CNT are processed into fibers or films, and this macroscopic CNT material is then embedded into a polymer matrix [65]. The four major fiber-spinning methods (Figure 4.15) used for polymer/CNT composites from both the solution and melt include dry-spinning [66], wet-spinning [67], dry-jet wet spinning (gel-spinning), and electrospinning [68]. An ancient solid-state spinning approach has been used for fabricating 100per cent CNT fibers from both forests and aerogels. Irrespective of the processing technique, in order to develop high-quality fibers many parameters need to be well controlled.

All spinning procedures generally involve:
(i) Fiber formation
(ii) coagulation/gelation/solidification and
(iii) drawing/alignment.

For all of these processes, the even dispersion of the CNT within the polymer solution or melt is very important. However, in terms of achieving excellent axial mechanical properties, alignment and orientation of the polymer chains and the CNT in the composite is necessary. Fiber alignment is accomplished in post-processing such as drawing/annealing and is key to increasing crystallinity, tensile strength, and stiffness [69].

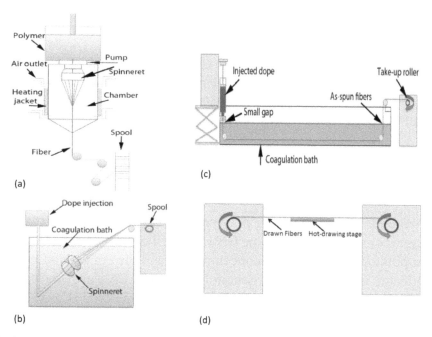

FIGURE 4.15   Schematics representing the various fiber processing methods (a) dry-spinning; (b) wet-spinning; (c) dry-jet wet or gel-spinning; and (d) post-processing by hot-stage drawing.

4.5   COMPUTATIONAL METHODS

Computational approaches to obtain solubility and diffusion coefficients of small molecules in polymers have focused primarily upon equilibrium molecular dynamics (MD) and Monte Carlo (MC) methods. These have been thoroughly reviewed by several investigators [70–71].

Computational approach can play an important role in the development of the CNT-based composites by providing simulation results to help on the understanding, analysis and design of such nanocomposites. At the nanoscale, analytical models are difficult to establish or too complicated to solve, and tests are extremely difficult and expensive to conduct. Modeling and simulations of nanocomposites, on the other hand, can be achieved readily and cost effectively on even a desktop computer. Characterizing the mechanical properties of CNT-based composites is just one of the many important and urgent tasks that simulations can follow out [72].

Computer simulations on model systems have in recent years provided much valuable information on the thermodynamic, structural and transport properties of classical dense fluids. The success of these methods rests primarily on the fact that a model containing a relatively small number of particles is in general found to be sufficient to simulate the behavior of a macroscopic system. Two distinct techniques of computer simulation have been developed which are known as the method of molecular dynamics and the Monte Carlo method [73–75].

Instead of adopting a trial-and-error approach to membrane development, it is far more efficient to have a real understanding of the separation phenomena to guide membrane design [76–79]. Similarly, methods such as MC, MD and other computational techniques have improved the understanding of the relationships between membrane characteristics and separation properties. In addition to these inputs, it is also beneficial to have simple models and theories that give an overall insight into separation performance [80–83].

4.5.1 PERMEANCE AND SELECTIVITY OF SEPARATION MEMBRANES

A membrane separates one component from another on the basis of size, shape or chemical affinity. Two characteristics dictate membrane performance, permeability, that is the flux of the membrane, and selectivity or the membrane's preference to pass one species and not another[84].

A membrane can be defined as a selective barrier between two phases, the "selective" being inherent to a membrane or a membrane processes. The membrane separation technology is proving to be one of the most significant unit operations. The technology inherits certain advantages over other methods. These advantages include compactness and light weight, low labor intensity, modular design that allows for easy expansion or operation at partial capacity, low maintenance, low energy requirements, low cost, and environmentally friendly operations. A schematic representation of a simple separation membrane process is shown in Figure 4.16.

A feed stream of mixed components enters a membrane unit where it is separated into a retentate and permeate stream. The retentate stream is typically the purified product stream and the permeate stream contains the waste component.

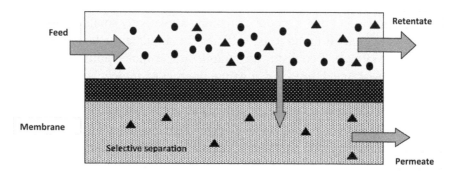

FIGURE 4.16 Schematic of membrane separation.

A quantitative measure of transport is the flux (or permeation rate), which is defined as the number of molecules that pass through a unit area per unit time [85]. It is believed that this molecular flux follows Fick's first law. The flux is proportional to the concentration gradient through the membrane. There is a movement from regions of high concentration to regions of low concentration, which may be expressed in the form:

$$J = -D\frac{dc}{dx} \tag{4.5}$$

By assuming a linear concentration gradient across the membrane, the flux can be approximated as:

$$J = -D\frac{C_2 - C_1}{L} \tag{4.6}$$

Where $C_1 = c(0)$ and $C_2 = c(L)$ are the downstream and upstream concentrations (corresponding to the pressures p_1 and p_2 via sorption isotherm $c(p)$, respectively, and L is the membrane thickness, as labeled in Figure 4.17.

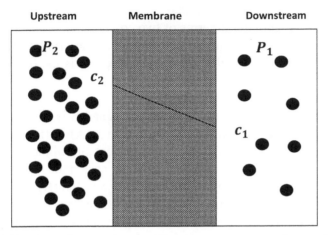

FIGURE 4.17 Separation membrane with a constant concentration gradient across membrane thickness L. The membrane performance of various materials is commonly compared using the thickness, independent material property, the permeability, which is related to the flux as:

$$P = \frac{JL}{P_2 - P_1} = \left(\frac{C_2 - C_1}{P_2 - P_1} \right) D \qquad (4.7)$$

In the case where the upstream pressure is much greater than the downstream pressure ($p_2 >> p_1$ and $C_2 >> C_1$) the permeability can be simplified so:

$$P = \frac{C_2}{P_2} D \qquad (4.8)$$

The permeability is more commonly used to describe the performance of a membrane than flux. This is because the permeability of a homogenous stable membrane material is constant regardless of the pressure differential or membrane thickness and hence it is easier to compare membranes made from different materials.

By introducing a solubility coefficient, the ratio of concentration over pressure C_2 / p_2, when sorption isotherm can be represented by the Henry's law, the permeability coefficient may be expressed simply as:

$$P = SD \qquad (4.9)$$

This form is useful as it facilitates the understanding of this physical property by representing it in terms of two components:

Solubility which is an equilibrium component describing the concentration of gas molecules within the membrane, that is the driving force, and Diffusivity, which is a dynamic component describing the mobility of the gas molecules within the membrane.

The separation of a mixture of molecules A and B is characterized by the selectivity or ideal separation factor $\alpha_{A/B} = P\,(A)/\,P\,(B)$, the ratio of permeability of the molecule. A over the permeability of the molecule B. According to Eq. (4.9), it is possible to make separations by diffusivity selectivity $D\,(A)/\,D\,(B)$ or solubility selectivity $S\,(A)/\,S\,(B)$ [85–86]. This formalism is known in membrane science as the solution-diffusion mechanism. Since the limiting stage of the mass transfer is overcoming of the diffusion energy barrier, this mechanism implies the activated diffusion. Because of this, the temperature dependences of the diffusion coefficients and permeability coefficients are described by the Arrhenius equations.

Gas molecules that encounter geometric constrictions experience an energy barrier such that sufficient kinetic energy of the diffusing molecule or the groups that form this barrier, in the membrane is required in order to overcome the barrier and make a successful diffusive jump. The common form of the Arrhenius dependence for the diffusion coefficient can be expressed as:

$$D_A = D_A^* \exp(-\Delta E_a\,/\,RT) \qquad\qquad (4.10)$$

For the solubility coefficient the Van't Hoff equation holds:

$$S_A = S_A^* \exp(-\Delta H_a\,/\,RT) \qquad\qquad (4.11)$$

Where $\Delta H_a < 0$ is the enthalpy of sorption. From Eq. (4.9), it can be written:

$$P_A = P_A^* \exp(-\Delta E_p\,/\,RT) \qquad\qquad (4.12)$$

Where $\Delta E_p = \Delta E_a + \Delta H_a$ are known to diffuse within nonporous or porous membranes according to various transport mechanisms. Table 4.3 illustrates the mechanism of transport depending on the size of pores. For very narrow pores, size sieving mechanism is realized that can be considered as a case of activated diffusion. This mechanism of diffusion is most common in the case of extensively studied nonporous polymeric membranes. For wider pores, the surface diffusion (also an activated diffusion process) and the Knudsen diffusion are observed [87–89].

TABLE 4.3 Transport mechanisms

Mechanism	Schamatic	Process
Activated diffusion		Constriction energy barrier ΔE_a
Surface diffusion		Adsorption – site energy barrier ΔE_s
Knudsen diffusion		Direction and velocity $\bar{d}$ $\bar{u}$

Sorption does not necessarily follow Henry's law. For a glassy polymer an assumption is made that there are small cavities in the polymer and the sorption at the cavities follows Langmuir's law. Then, the concentration in the membrane is given as the sum of Henry's law adsorption and Langmuir's law adsorption

$$C = K_P P + \frac{C_h^* b_P}{1 + b_P} \tag{4.13}$$

It should be noted that the applicability of solution (sorption)-diffusion model has nothing to do with the presence or absence of the pore.

4.5.2 DIFFUSIVITY

The diffusivity through membranes can be calculated using the time-lag method [90]. A plot of the flow through the membrane versus time reveals an initial transient permeation followed by steady state permeation. Extending the linear section of the plot back to the intersection of the x axis gives the value of the time-lag (θ) as shown in Figure 4.18.

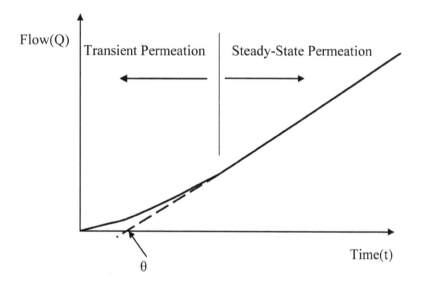

FIGURE 4.18 Calculation of the diffusion coefficient using the time-lag method, once the gradient is constant and steady state flow through the membrane has been reached, a extrapolation of the steady state flow line back to the x-axis where the flow is 0reveals the value of the time-lag (θ). The time lag relates to the time it takes for the first molecules to travel through the membrane and is thus related to the diffusivity. The diffusion coefficient can be calculated from the time-lag and the membrane thickness as shown in Eq.(4.14) [91–92].

$$D = \frac{\Delta x}{6\theta} \qquad (4.14)$$

Surface diffusion is the diffusion mechanism which dominates in the pore size region between activation diffusion and Knudsen diffusion [93].

4.5.3 SURFACE DIFFUSION

A model that well described the surface diffusion on the pore walls was proposed many years ago. It was shown to be consistent with transport parameters in porous polymeric membranes. When the pore size decreased below a certain level, which depends on both membrane material and the permeability coefficient exceeds the value for free molecular flow (Knudsen diffusion), especially in the case of organic vapors. Note that surface diffusion usually occurs simultaneously with Knudsen diffusion but it is the dominant mechanism within a certain pore size. Since surface diffusion is also a form of activated diffusion, the energy barrier is the energy required for the molecule to jump from one adsorption site to another across the surface of the pore. By allowing the energy barrier to be proportionate to the enthalpy of adsorption, Gilliland *et al*, established an equation for the surface-diffusion coefficient expressed here as: [94]:

$$D_S = D_S^* \exp(\frac{-aq}{RT}) \qquad (4.15)$$

Where D_S^* is a pre-exponential factor depending on the frequency of vibration of theadsorbed molecule normal to the surface and the distance from one adsorption site to the next. The quantity the heat of adsorption is $(q > 0)$ and a proportionality constant is $(0 < a < 1)$. The energy barrier separates the adjacent adsorption sites. An important observation is that more strongly adsorbed molecules are less mobile than weakly adsorbed molecules [95].

In the case of surface diffusion, the concentration is well described by Henry's law $c = Kp$, where K is $K = K_0 \exp(q/RT)$ [95–96]. Since solubility is the ratio of the equilibrium concentration over pressure, the solubility is equivalent to the Henry's law coefficient.

$$S_S = K_0 \, exp\left(q|RT\right) \qquad (4.16)$$

Which implies the solubility is a decreasing function of temperature. The product of diffusivity and solubility gives:

$$P_S = P_S^* \, exp\left(\frac{(1-a)q}{RT}\right) \qquad (4.17)$$

Since $0 < a < 1$ the total permeability will decrease with increased temperature meaning that any increase in the diffusivity is counteracted by a decrease in surface concentration[95].

4.5.4 KNUDSEN DIFFUSION

Knudsen diffusion[95, 97–99] depending on pressure and mean free path which applies to pores between 10 Å and 500 Å in size [100]. In this region, the mean free path of molecules is much larger than the pore diameter. It is common to use Knudsen number $K_n = \lambda/d$ to characterize the regime of permeation through pores. When $K_n \ll 1$, viscous (Poiseuille) flow is realized. The condition for Knudsen diffusion is $K n \gg 1$. An intermediate regime is realized when $K_n \approx 1$. The Knudsen diffusion coefficient can be expressed in the following form:

$$D_K = \frac{d}{3\tau}\bar{u} \tag{4.18}$$

This expression shows that the separation outcome should depend on the differences in molecular speed (or molecular mass). The average molecular speed is calculated using the Maxwell speed distribution as:

$$\bar{u} = \sqrt{\frac{8RT}{\pi m}} \tag{4.19}$$

And the diffusion coefficient can be presented as:

$$D_K = \left(\frac{d}{3\tau}\right)(\frac{8RT}{\pi m})^{1/2} \tag{4.20}$$

For the flux in the Knudsen regime the following equation holds [101–102]:

$$J = n\pi d^2 \Delta p D_K / 4RTL \tag{4.21}$$

After substituting Eqs. (4.20) into (4.21), one has the following expressions for the flux and permeability coefficient is:

$$J = \left(\frac{n\pi^{\frac{1}{2}} d^3 \Delta p}{6\tau L} \right) \left(\frac{2}{mRT} \right)^{1/2} \tag{4.22}$$

$$P = \left(\frac{n\pi^{\frac{1}{2}} d^3}{6\tau} \right) \left(\frac{2}{mRT} \right)^{1/2} \tag{4.23}$$

Two important conclusions can be made from analysis of Eqs. (4.22) and (4.23). First, selectivity of separation in Knudsen regime is characterized by the ratio $\alpha_{ij} = (M_j/M_i)^{1/2}$. It means that membranes where Knudsen diffusion predominates are poorly selective.

The most common approach to obtain diffusion coefficients is equilibrium molecular dynamics. The diffusion coefficient that is obtained is a self-diffusion coefficient. Transport-related diffusion coefficients are less frequently studied by simulation but several approaches using non-equilibrium MD (NEMD) simulation can be used.

4.5.5 MOLECULAR DYNAMICS (MD) SIMULATIONS

Conducting experiments for material characterization of the nanocomposites is a very time consuming, expensive and difficult. Many researchers are now concentrating on developing both analytical and computational simulations. MD simulations are widely being used in modeling and solving problems based on quantum mechanics. Using Molecular dynamics it is possible to study the reactions, load transfer between atoms and molecules. If the objective of the simulation is to study the overall behavior of CNT-based composites and structures, such as deformations, load and heat-transfer mechanisms then the continuum mechanics approach can be applied safely to study the problem effectively [103].

MD tracks the temporal evolution of a microscopic model system by integrating the equations of motion for all microscopic degrees of freedom. Numerical integration algorithms for initial value problems are used for this purpose, and their strengths and weaknesses have been discussed in simulation texts [104–106].

MD is a computational technique in which a time evolution of a set of interacting atoms is followed by integrating their equations of motion. The forces between atoms are due to the interactions with the other atoms. A trajectory is calculated in a 6-N dimensional phase space (three position and three momentum components for each of the N atoms). Typical MD simulations of CNT composites are performed on molecular systems containing up to tens of thousands of atoms and for simulation times up to nanoseconds. The physical quantities of the system are represented by averages over configurations distributed according to the chosen statistical ensemble. A trajectory obtained with MD provides such a set of configurations. Therefore the computation of a physical quantity is obtained as an arithmetic average of the instantaneous values. Statistical mechanics is the link between the nanometer behavior and thermodynamics. Thus the atomic system is expected to behave differently for different pressures and temperatures[107].

The interactions of the particular atom types are described by the total potential energy of the system, U, as a function of the positions of the individual atoms at a particular instant in time

$$U = U\left(X_i, \ldots, X_n\right) \tag{4.24}$$

Where X_1 represents the coordinates of atom i in a system of N atoms. The potential equation is invariant to the coordinate transformations, and is expressed in terms of the relative positions of the atoms with respect to each other, rather than from absolute coordinates [107].

MD is readily applicable to a wide range of models, with and without constraints. It has been extended from the original microcanonical ensemble formulation to a variety of statistical mechanical ensembles. It is flexible and valuable for extracting dynamical information. The Achilles' heel of MD is its high demand of computer time, as a result of which the longest times that can be simulated with MD fall short of the longest relaxation times of most real-life macromolecular systems by several orders of magnitude. This has two important consequences. (a) Equilibrating an atomistic model polymer system with MD alone is problematic; if one starts from an improbable configuration, the simulation will not have the time to depart significantly from that configuration and visit the regions of phase space that contribute most significantly to the properties. (b) Dynamical processes with characteristic times longer than approximately 10^{-7} scannot be probed directly; the relevant correlation functions do not decay to zero within the simulation time and thus their long-time tails are inaccessible, unless some extrapolation is invoked based on their short-time behavior.

Recently, rigorous multiple time step algorithms have been invented, which can significantly augment the ratio of simulated time to CPU time. Such an algorithm is the reversible Reference System Propagator Algorithm (rRESPA) [108–109]. This algorithm invokes a Trotter factorization of the Liouville operator in the numerical integration of the equations of motion: fast-varying (e.g., bond stretching and bond angle bending) forces are updated with a short-time step Δt, while slowly varying forces (e.g., nonbonded interactions, which are typically expensive to calculate, are updated with a longer time step Δt. Using $\delta t = 1 fs$ and $\Delta t = 5 ps$, one can simulate 300 ns of real time of a polyethylene melt ona modest workstation[110]. This is sufficient for the full relaxation of a system of C_{250} chains, but not of longer-chain systems.

A paper of Furukawa and Nitta is cited first to understand the NEMD simulation semi-quantitatively, since, even though the paper deals with various pore shapes, complicated simulation procedure is described clearly.

MD simulation is more preferable to study the non-equilibrium transport properties. Recently some NEMD methods have also been developed, such as the grand canonical molecular dynamics (GCMD) method [111–112] and the dual control volume GCMD technique (DCV- GCMD) [113–114]. These methods provide a valuable clue to insight into the transport and separation of fluids through a porous medium. The GCMD method has recently been used to investigate pressure-driven and chemical potential-driven gas transport through porous inorganic membrane [115].

4.5.5.1   EQUILIBRIUM MD SIMULATION

A self-diffusion coefficient can be obtained from the mean-square displacement (MSD) of one molecule by means of the Einstein equation in the form [115]:

$$D_A^* = \frac{1}{6N_\alpha} \lim_{t \to \infty} \frac{d}{dt} \left(r_i(t) - r_i(0) \right)^2 \tag{4.25}$$

Where Na is the number of molecules, $r_i(t)$ and $r_i(0)$ are the initial and final (at time t) positions of the center of mass of one molecule i over the time interval t, and $(r_i(t)-r_i(0))^2$ is MSD averaged over the ensemble. The Einstein relationship assumes a random walk for the diffusing species. For slow diffusing species, anomalous diffusion is sometimes observed and is characterized by:

$$\left(r_i(t) - r_i(0) \right)^2 \propto t^n \tag{4.26}$$

Where n < 1 ($n = 1$ for the Einstein diffusion regime). At very short times ($t < 1$ ps), the MSD may be quadratic iv n time ($n = 2$) which is characteristic of 'free flight' as may occur in a pore or solvent cage prior to collision with the pore or cage wall. The result of anomalous diffusion, which may or may not occur in intermediate time scales, is to create a smaller slope at short times, resulting in a larger value for the diffusion coefficient. At sufficiently long times (the hydrodynamic limit), a transition from anomalous to Einstein diffusion ($n = 1$) may be observed[71].

An alternative approach to MSD analysis makes use of the center-of-mass velocity autocorrelation function (VACF) or Green–Kubo relation, given as follows [116]:

$$D = \frac{1}{3} \int (v_i(t).v_i(0)) dt \tag{4.27}$$

Concentration in the simulation cell is extremely low and its diffusion coefficient is an order of magnitude larger than that of the polymeric segments. Under these circumstances, the self-diffusion and mutual diffusion coefficients of the penetrant are approximately equal, as related by the Darken equation in the following form:

$$D_{AB} = (D_A^* x_B + D_B^* x_A) \left(\frac{d \ln f_A}{d \ln c_A} \right) \tag{4.28}$$

In the limit of low concentration of diffusion $x_A \approx 0$, Eq. (4.28) reduces to:

$$D_A^* \equiv D_{AB} \qquad (4.29)$$

4.5.5.2 NON-EQUILIBRIUM MD SIMULATION

Experimental diffusion coefficients, as obtained from time-lag measurements, report a transport diffusion coefficient which cannot be obtained from equilibrium MD simulation. Comparisons made in the simulation literature are typically between time-lag diffusion coefficients (even calculated for glassy polymers without correction for dual-mode contributions and self-diffusion coefficients. As discussed above, mutual diffusion coefficients can be obtained directly from equilibrium MD simulation but simulation of transport diffusion coefficients require the use of NEMD methods, that are less commonly available and more computationally expensive[117].

For these reasons, they have not been frequently used. One successful approach is to simulate a chemical potential gradient and combine MD with GCMC methods (GCMC–MD), as developed by Heffelfinger and coworkers [114] and MacElroy [118]. This approach has been used to simulate permeation of a variety of small molecules through nanoporous carbon membranes, carbon nanotubes, porous silica and self-assembled monolayers [119–121]. A diffusion coefficient then can be obtained from the relation:

$$D = \frac{KT}{F}(V) \qquad (4.30)$$

4.5.6 GRAND CANONICAL MONTE CARLO (GCMC) SIMULATION

A standard GCMC simulation is employed in the equilibrium study, while MD simulation is more preferable to study the non-equilibrium transport properties [104].

Monte Carlo method is formally defined by the following quote as: Numerical methods that are known as Monte Carlo methods can be loosely described as statistical simulation methods, where statistical simulation is defined in quite general terms to be any method that utilizes sequences of random numbers to perform the simulation [122].

The name "Monte Carlo" was chosen because of the extensive use of random numbers in the calculations [104]. One of the better known applications of Monte Carlo simulations consists of the evaluation of integrals by generating suitable random numbers that will fall within the area of integration. A simple example of how a MC simulation method is applied to evaluate the value of π is illustrated in Figure 4.19. By considering a square that inscribes a circleof a diameter R, one can deduce that the area of the square is R^2,and the circle has anarea of $\pi R^2/4$. Thus, the relative area of the circle and the square will be $\pi/4$. A large number of two independent random numbers (with x and y coordinates) of trial shots is generated within the square to determine whether each of them falls inside of the circle or not. After thousands or millions of trial shots, the computer program keeps counting the total number of trial shots inside the square and the number of shots landing inside the circle. Finally, the value of $\pi/4$ can be approximated based on the ratio of the number of shots that fall inside the circle to the total number of trial shots.

As stated earlier, the value of an integral can be calculated via MC methods by generating a large number of random points in the domain of that integral. Equation (4.31) shows a definite integral:

$$F = \int_a^b f(x)\,dx \tag{4.31}$$

Where $f(x)$ is a continuous and real-valued function in the interval $[a, b]$. The integral can be rewritten as [104]:

$$F = \int_a^b dx \left(\frac{f(x)}{\rho(x)} \right) \rho(x) \cong \frac{f(\xi_i)}{\rho(\xi_i)}_\tau \tag{4.32}$$

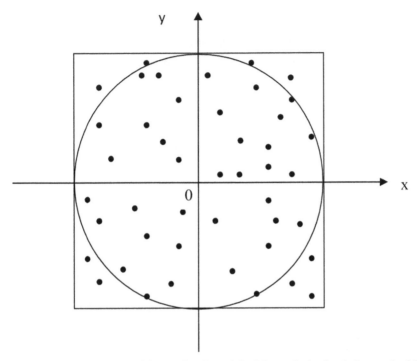

FIGURE 4.19   Illustration of the application of the Monte Carlo simulation method for the calculation of the value of π by generating a number of trial shots, in which the ratio of the number of shots inside the circle to the total number of trial shots will approximately approach the ratio of the area of the circle to the area of the square.

If the probability function is chosen to be a continuous uniform distribution, then:

$$p(x) = \frac{1}{(b-a)} \, a \le x \le b \tag{4.33}$$

Subsequently, the integral, F, can be approximated as:

$$F \approx \frac{(b-a)}{\tau} \sum_{i=1}^{\tau} f(\xi_i) \tag{4.34}$$

In a similar way to the MC integration methods, MC molecular simulation methods rely on the fact that a physical system can be defined to possess a definite energy distribution function, which can be used to calculate thermodynamic properties.

The applications of MC are diverse such as Nuclear reactor simulation, Quantum chromo dynamics, Radiation cancer therapy, Traffic flow, Stellar evolution, Econometrics, Dow Jones forecasting, Oil well exploration, VSLI design [122]

The MC procedure requires the generation of a series of configurations of the particles of the model in a way which ensures that the configurations are distributed in phases pace according to some prescribed probability density.

The mean value of any configurational property determined from a sufficiently large number of configurations provides an estimate of the ensemble-average value of that quantity; the nature of the ensemble average depends upon the chosen probability density.

These machine calculations provide what is essentially exact information on the consequences of a given intermolecular force law. Application has been made to hard spheres and hard disks, to particles interacting through a Lennard-Jones 12–6 potential function and other continuous potentials of interest in the study of simple fluids, to systems of charged particles [123].

The MC technique is a stochastic simulation method designed to generate a long sequence, or 'Markov chain' of configurations that asymptotically sample the probability density of an equilibrium ensemble of statistical mechanics [105, 116]. For example, a MC simulation in the canonical (NVT) ensemble, carried out under the macroscopic constraints of a prescribed number of molecules N, total volume V and temperature T, samples configurations r_p with probability proportional to $\exp[-\beta v(r_p)]$, with $\beta = 1/(k_B T)$, k_B being the Boltzmann constant and T the absolute temperature. Thermodynamic properties are computed as averages over all sampled configurations.

The efficiency of a MC algorithm depends on the elementary moves it employs to go from one configuration to the next in the sequence. An attempted move typically involves changing a small number of degrees of freedom; it is accepted or rejected according to selection criteria designed so that the sequence ultimately conforms to the probability distribution of interest. In addition to usual moves of molecule translation and rotation practiced for small-molecule fluids, special moves have been invented for polymers. The reptation (slithering snake) move for polymer chains involves deleting a terminal segment on one end of the chain and appending a terminal segment on the other end, with the newly created torsion angle being assigned a randomly chosen value[124].

In most MC algorithms the overall probability of transition from some state (configuration) m to some other state n, as dictated by both the attempt and the selection stages of the moves, equals the overall probability of transition from n to m; this is the principle of detailed balance or 'microscopic reversibility'. The probability of attempting a move from state m to state n may or may not be equal to that of attempting the inverse move from state n to state m. These probabilities of attempt are typically unequal in 'bias' MC algorithms, which incorporate information about the system energetics in attempting moves. In bias MC, detailed balance is ensured by appropriate design of the selection criterion, which muster move the bias inherent in the attempt [105, 116].

4.5.7   MEMBRANE MODEL AND SIMULATION BOX

The MD simulations [125] can be applied for the permeation of pure and mixed gases across carbon membranes with three different pore shapes: the diamond pore (DP), zigzag path (ZP) and straight path (SP), each composed of micro-graphite crystalline. Three different pore shapes can be considered: DP, ZP, and SP.

The simulation box is divided into three regions where the chemical potential for each component is the same. The middle region (M-region) represents the membrane with slit pores, in which the distances between the two adjacent carbon atoms (Lcc) and two adjacent graphite basal planes (Δ) .

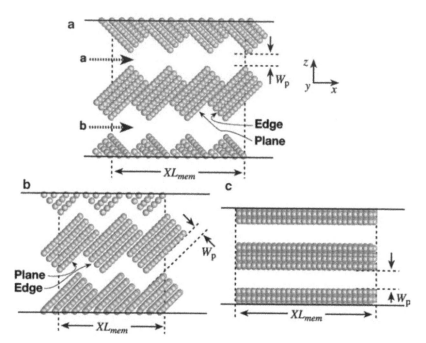

FIGURE 4.20 Three membrane pore shapes; (a) diamond path (DP), (b) zigzag path (ZP), (c) straight path (SP). Figure 4.20(a)–4.20(c) shows the cross-sectional view of each pore shape. DP (A) has two different pore mouths; one a large (pore a) and the other a small mouth (pore b). ZP (B) has zigzag shaped pores whose sizes (diameters) are all the same at the pore entry. SP (C) has straight pores which can be called slit-shaped pores. In a simulation system, we investigate the equilibrium selective adsorption and non-equilibrium transport and separation of gas mixture in the nanoporous carbon membrane are modeled as slits from the layer structure of graphite. A schematic representation of the system used in our simulations is shown in Figure s 4.21(a) and 4.21(b), in which the origin of the coordinates is at the center of simulation box and transport takes place along the x-direction in the non-equilibrium simulations. In the equilibrium simulations, the box as shown in Figure 4.21(a) is employed, whose size is set as 85.20 nm × 4.92 nm × (1.675 + W) nm in x-, y-, and z-directions, respectively, where W is the pore width, (i.e., the separation distance between the centers of carbon atoms on the two layers forming a slit pore) (Figure 4.21). L_{cc} is the separation distance between two centers of adjacent carbon atom; L_m is the pore length; W is the pore width, Δ is the separation distance between two carbon atom centers of two adjacent layers [126] .

Period boundary conditions are employed in all three directions. In the non-equilibrium molecular dynamics simulations in order to use period boundary conditions in three directions, we have to divide the system into five regions as shown in Figure 4.21(b).

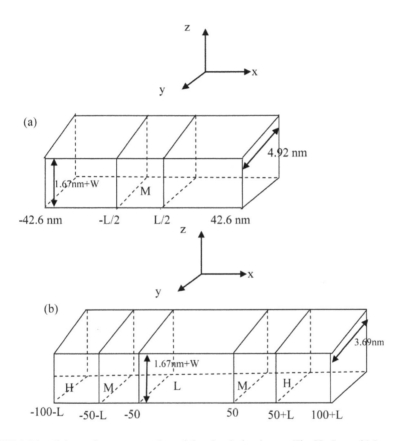

FIGURE 4.21 Schematic representation of the simulation boxes. The H-, L- and M-areas correspond to the high and low chemical potential control volumes, and membrane, respectively. Transport takes place along the x-direction in the non-equilibrium simulations. (a) Equilibrium adsorption simulations and (b) non-equilibrium transport simulations. L is the membrane thickness and W is the pore width.

The difference in the gas density between the H- and L- region is the driving force for the gas permeation through the M-region which represents the membrane (Figure 4.22).

The transition and rotational velocities are given to each inserted molecules randomly based on the Gaussian distribution around an average velocity corresponding to the specified temperature.

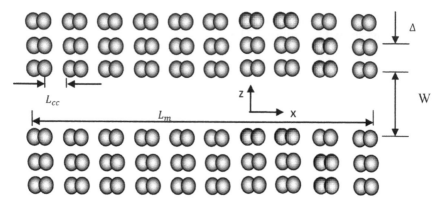

FIGURE 4.22 Schematic representation of slit pore. Each symmetric box has three regions. Two are density control; H-region (high density) and L-region (low density) and one is free of control M-region which is placed between the H- and L- region. For each simulation, the density in the H-region, ρ_H, is maintained to be that of the feed gas and the density in the L-region is maintained at zero, corresponding to the vacuum.

Molecules spontaneously move from H- to L- region via leap-frog algorithm and a non-equilibrium steady state is obtained at the M-region. During a simulation run, equilibrium with the bulk mass at the feed side at the specified pressure and temperature is maintained at the H-region by carrying out GCMC creations and destructions in terms of the usual acceptance criteria [28]. Molecules entered the L-region were moved out immediately to keep vacuum. The velocities of newly inserted molecules were set to certain values in terms of the specified temperature by use of random numbers on the Gaussian distribution.

4.6   CONCLUDING REMARKS

The concept of membrane processes is relatively simple but nevertheless often unknown. Membrane separation processes can be used for a wide range of applications. The separation mechanism in MF/UF/NF is mainly the size exclusion, which is indicated in the nominal ratings of the membranes. The other separation mechanism includes the electrostatic interactions between solutes and membranes, which depends on the surface and physiochemical properties of solutes and membranes. The available range of membrane materials includes polymeric, carbon, silica, zeolite and other ceramics, as well as composites. Each type of membrane can have a different porous structure. Nowadays, there are more reports on the fluid transport through porous CNTs/polymer membrane. Computational approach can play an important role in the development of the CNT-based composites by providing simulation results to help on the understanding, analysis and design of such nanocomposites. Computational approaches to obtain solubility and diffusion coefficients of small molecules in polymers have focused primarily upon molecular dynamics and Monte Carlo methods. Molecular dynamics simulations are widely being used in modeling and solving problems based on quantum mechanics. Using Molecular dynamics it is possible to study the reactions, load transfer between atoms and molecules. Monte Carlo molecular simulation methods rely on the fact that a physical system can be defined to possess a definite energy distribution function, which can be used to calculate thermodynamic properties. The Monte Carlo technique is a stochastic simulation method designed to generate a long sequence, or 'Markov chain' of configurations that asymptotically sample the probability density of an equilibrium ensemble of statistical mechanics. So using from molecular dynamic or Monte Carlo techniques can be useful to simulate the membrane separation process depends on the purpose and the condition of process. Symbol

Symbol	Definition
r_p	Pore radius
Γ	Surface tension
T	Thickness of the adsorbate film
V_L	Molecular volume of the condensate
a_1 and a_2	Unit vectors
G_1, G_2	The material coordinates of a point in the initial configuration
e_1, e_2 and e_3	The coordinates in the current configuration
R	The radius of the modeled SWCNT
J	Molecular flux
D	The diffusivity (diffusion coefficient)
$c(x)$	Concentration
X	Position across the membrane

P	Permeability
S	Solubility coefficient
K_p	Henry's law constant
B	Hole affinity constant
C_h^*	Saturation constant
Q	Flow
T	Time
Θ	Time-lag
Q	Heat of adsorption
A	Proportionality constant
aq	Energy barrier
K	Temperature dependent Henry's law coefficient
K_0	Proportionality constant
P_s^*	Constant
Λ	The mean free path of molecules
D	Pore diameter
K_n	Knudsen number
T	Pore tortuosity
$\bar{u}$	Average molecular speed
M	Molecular mass
N	Surface concentration of pores
Δp	Pressure drop across the membrane
L	Membrane thickness
D_A^*	Self-diffusion coefficient
Na	The number of molecules
f_A	Fugacity
c_A	The concentration of diffusant A
F	Applied external force
V	The center-of-mass velocity component

$\rho(x)$	An arbitrary probability distribution function
ξ_i	The random numbers generated for each trial
T	The number of trials
RO	Reverse osmosis
NF	Nanofiltration
UF	Ultrafiltration
MF	Microfiltration
MWCO	Molecular Weight Cut-Off
IUPAC	International Union of Pure and Applied Chemistry
SWNT	Single-Walled Carbon Nanotube
MWNT	Multiwalled Carbon Nanotube
CNT	Carbon nanotube
AFM	Atomic Force Microscopy
MD	Molecular Dynamics
MC	Monte Carlo
NEMD	Nonequilibrium MD
RESPA	Reference System Propagator Algorithm
GCMD	Grand Canonical Molecular Dynamics
DCV- GCMD	Dual-volume GCMD
MSD	Mean-Square Displacement
DP	Diamond Pore
ZP	Zigzag Path
SP	Straight Path

KEYWORDS

- **Computational methods**
- **Filtration**
- **Membrane**
- **Membrane types**

REFERENCES

1. Majeed, S.; et al. Multi-walled carbon nanotubes (MWCNTs) mixed polyacrylonitrile (PAN) ultrafiltration membranes. *J. Membrane Sci.* **2012,** *403,* 101–109.
2. Macedonio, F.; and Drioli, E.; Pressure-driven membrane operations and membrane distillation technology integration for water purification. *Desalination.* **2008,** *223(1),* 396–409.
3. Merdaw, A. A.; Sharif, A. O.; and Derwish, G. A. W.; Mass transfer in pressure-driven membrane separation processes, part II. *Chem.Eng. J.* **2011,** *168(1),* 229–240.
4. Van Der Bruggen, B.; et al. A review of pressure-driven membrane processes in wastewater treatment and drinking water production.*Environ. Progress.* **2003,** *22(1),* 46–56.
5. Cui, Z. F.; and Muralidhara, H.S.; Membrane Technology: A Practical Guide to Membrane Technology and Applications in Food and Bioprocessing. Elsevier; **2010,** 288.
6. Shirazi, S.; Lin, C. J.; and Chen, D.; Inorganic fouling of pressure-driven membrane processes—a critical review. *Desalination.* **2010,** *250(1),* 236–248.
7. Pendergast, M. M.; and Hoek, E. M. V.; A review of water treatment membrane nanotechnologies. *Energy Environ. Sci.* **2011,** *4(6),* 1946–1971.
8. Hilal, N.; et al. A comprehensive review of nanofiltration membranes: Treatment, pretreatment, modelling, and atomic force microscopy. *Desalination.* **2004,** *170(3),* 281–308.
9. Srivastava, A.; Srivastava, S.; and Kalaga, K.; Carbon nanotube membrane filters. In:Springer Handbook of Nanomaterials. Springer; **2013,** 1099–1116.
10. Colombo, L.; and Fasolino, A. L.; Computer-Based Modeling of Novel Carbon Systems and their Properties: Beyond Nanotubes. Springer; *3,* **2010,** 258.
11. Polarz, S.; and Smarsly, B.; Nanoporous Materials. *J. Nanosci. Nanotechnol.***2002,** *2(6),*581–612.
12. Gray-Weale, A. A.; et al. Transition-state theory model for the diffusion coefficients of small penetrants in glassy polymers. *Macromolecule.***1997,** *30(23),* 7296–7306.
13. Rigby, D.; and Roe, R.; Molecular dynamics simulation of polymer liquid and glass. I. glass transition. *J. Chem. Phys.***1987,** *87,* 7285.
14. Freeman, B. D.; Yampolskii,Y. P.; and Pinnau, I.; Materials Science of Membranes for Gas and Vapor Separation. Wiley. com. **2006,** 466.
15. Hofmann, D.; et al. Molecular modeling investigation of free volume distributions in stiff chain polymers with conventional and ultrahigh free volume: comparison between molecular modeling and positron lifetime studies.*Macromolecule.* **2003,** *36(22),* 8528–8538.
16. Greenfield, M. L.; and Theodorou, D. N.; Geometric analysis of diffusion pathways in glassy and melt atactic polypropylene.*Macromolecule.* **1993,** *26(20),* 5461–5472.
17. Baker, R. W.; Membrane Technology and Applications. John Wiley & Sons; **2012,** 59218. Strathmann, H.; Giorno, L.; and Drioli, E.; Introduction to Membrane Science and Technology. Wiley-VCH Verlag & Company; **2011,** 544.
19. Chen, J. P.; et al. Membrane Separation: Basics and Applications, in Membrane and Desalination Technologies. Ed.Wang, L. K.; et al. Humana Press;**2008,** 271–332.
20. Mortazavi, S.; Application of Membrane Separation Technology to Mitigation of Mine Effluent and Acidic Drainage. Natural Resources Canada; **2008,** 194.

21. Porter, M. C.; Handbook of Industrial Membrane Technology. Noyes Publications; **1990,** 604.
22. Naylor, T. V.; Polymer Membranes: Materials, Structures and Separation Performance. Rapra Technology Limited; **1996,** 136.
23. Freeman, B. D.; Introduction to membrane science and technology. by heinrich strathmann. *Angewandte Chemie Int. Edn.* **2012,** *51(38),* 9485–9485.
24. Kim, I.; Yoon, H.; and Lee, K. M.; Formation of integrally skinned asymmetric polyetherimide nanofiltration membranes by phase inversion process. *J. Appl. Polym. Sci.* **2002,** *84(6),* 1300–1307.
25. Khulbe, K. C.; Feng, C. Y.; and Matsuura,T.; Synthetic Polymeric Membranes: Characterization by Atomic Force Microscopy. Springer; **2007,** 198.
26. Loeb, L. B.; The Kinetic Theory of Gases. Courier Dover Publications; **2004,** 678.
27. Koros, W. J.; and Fleming, G. K.; Membrane-based gas separation. *J. Membrane Sci.* **1993,** *83(1),* 1–80.
28. Perry, J. D.; Nagai, K.; and Koros,W. J.; Polymer membranes for hydrogen separations. *MRS Bull.* **2006,** *31(10),* 745–749.
29. Yang, W.; et al. Carbon nanotubes for biological and biomedical applications. *Nanotechnology.* **2007,** *18(41),* 412001.
30. Bianco, A.; et al. Biomedical applications of functionalised carbon nanotubes. *Chem. Commun.* **2005,** *5,* 571–577.
31. Salvetat, J.; et al. Mechanical Properties of Carbon Nanotubes. Applied Physics A;**1999,** *69(3),* 255–260.
32. Zhang, X.; et al. Ultrastrong, stiff, and lightweight carbon-nanotube fibers. *Adv. Mater.* **2007,** *19(23),* 4198–4201.
33. Arroyo, M.; and Belytschko,T.; Finite crystal elasticity of carbon nanotubes based on the exponential cauchy-born rule.*Phys. Rev. B.* **2004,** *69(11),* 115415.
34. Wang, J.; et al. Energy and mechanical properties of single-walled carbon nanotubes predicted using the higher order cauchy-born rule. *Phys. Rev. B.***2006,** *73(11),* 115428.
35. Zhang, Y.; Single-Walled Carbon Nanotube Modelling Based on One-and Two-Dimensional Cosserat Continua. University of Nottingham; **2011.**
36. Wang, S.; Functionalization of Carbon Nanotubes: Characterization, Modeling and Composite Applications. Florida State University; **2006,** 193.
37. Lau, K.-t.; Gu, C.; and Hui, D.; A critical review on nanotube and nanotube/nanoclay related polymer composite materials. *Compos. Part B: Eng.* **2006,** *37(6),* 425–436.
38. Choi, W.; et al. Carbon nanotube-guided thermopower waves.*Mater. Today.* **2010,** *13(10),* 22–33.
39. Iijima, S.; Helical microtubules of graphitic carbon. *Nature.* **1991,** *354(6348),* 56–58.
40. Sholl, D. S.; and Johnson,J.; Making high-flux membranes with carbon nanotubes. *Sci.* **2006,** *312(5776),* 1003–1004.
41. Zang, J.; et al. Self-diffusion of water and simple alcohols in single-walled aluminosilicate nanotubes. *ACS Nano.* **2009,** *3(6),* 1548–1556.
42. Talapatra, S.; Krungleviciute, V.; and Migone, A. D.; Higher coverage gas adsorption on the surface of carbon nanotubes: evidence for a possible new phase in the second layer. *Phys. Rev. Lett.* **2002,***89(24),* 246106.
43. Pujari, S.; et al. Orientation dynamics in multiwalled carbon nanotube dispersions under shear flow. *J.Chem. Phys.* **2009,** *130,* 214903.

44. Singh, S.; and Kruse, P.; Carbon nanotube surface science. *Int. J. Nanotechnol.***2008**, *5(9)*, 900–929.
45. Baker, R.W.; Future directions of membrane gas separation technology.*Ind. Eng. Chem. Res.* **2002**, *41(6)*, 1393–1411.
46. Erucar, I.; and Keskin, S.; Screening metal–organic framework-based mixed-matrix membranes for CO2/CH4 separations. *Ind. Eng. Chem. Res.* **2011**, *50(22)*, 12606–12616.
47. Bethune, D. S.; et al. Cobalt-catalysed growth of carbon nanotubes with single-atomic-layer walls. *Nature.* **1993**, *363*, 605–607.
48. Iijima, S.; and Ichihashi,T.; Single-shell carbon nanotubes of 1-nm diameter.*Nature.* **1993**, *363*, 603–605.
49. Treacy, M.; Ebbesen, T.; and Gibson, J.; Exceptionally High Young's Modulus Observed for Individual Carbon Nanotubes. **1996**.
50. Wong, E. W.; Sheehan, P. E.; and Lieber, C.; Nanobeam mechanics: elasticity, strength, and toughness of nanorods and nanotubes. *Sci.* 1997, *277(5334)*, 1971–1975.
51. Thostenson, E. T.; Li,C.; and Chou, T. W.; Nanocomposites in context. *Compos. Sci. Technol.* **2005**, *65(3)*, 491–516.
52. Barski, M.; Kędziora, P.; and Chwał, M.; Carbon nanotube/polymer nanocomposites: a brief modeling overview. *Key Eng. Mater.* **2013**, *542*, 29–42.
53. Dresselhaus, M. S.; Dresselhaus, G.; and Eklund, P. C.; Science of Fullerenes and Carbon nanotubes:Ttheir Properties and Applications. Academic Press; **1996**, 965.
54. Yakobson, B.; and Smalley, R. E.; Some unusual new molecules—long, hollow fibers with tantalizing electronic and mechanical properties—have joined diamonds and graphite in the carbon family. *Am. Scientist.*1997, *85*, 324–337.
55. Guo, Y.; and Guo, W.; Mechanical and electrostatic properties of carbon nanotubes under tensile loading and electric field. *J. Phys. D: Appl. Phys.* **2003**, *36(7)*, 805.
56. Berger, C.; et al. Electronic confinement and coherence in patterned epitaxial graphene. *Sci.* **2006**, *312(5777)*, 1191–1196.
57. Song, K.; et al. Structural polymer-based carbon nanotube composite fibers: understanding the processing–structure–performance relationship.*Mater.* **2013**, *6(6)*, 2543–2577.
58. Park, O. K.; et al. Effect of surface treatment with potassium persulfate on dispersion stability of multi-walled carbon nanotubes. *Mater. Lett.* **2010**, *64(6)*, 718–721.
59. Banerjee, S.; Hemraj-Benny, T.; and Wong, S. S.; Covalent surface chemistry of single-walled carbon nanotubes. *A dv. Mater.* **2005**, *17(1)*, 17–29.
60. Balasubramanian, K.; and Burghard, M.; Chemically functionalized carbon nanotubes. *Small.* **2005**, *1(2)*, 180–192.
61. Xu, Z. L.; and Alsalhy Qusay, F.; Polyethersulfone (pes) hollow fiber ultrafiltration membranes prepared by PES/non-solvent/NMP solution. *J. Membrane Sci.* **2004**, *233(1–2)*, 101–111.
62. Chung, T. S.; Qin, J. J.; and Gu, J.; Effect of shear rate within the spinneret on morphology, separation performance and mechanical properties of ultrafiltration polyethersulfone hollow fiber membranes. *Chem. Eng. Sci.* **2000**, *55(6)*, 1077–1091.
63. Choi, J. H.; Jegal, J.; and Kim, W. N.; Modification of performances of various membranes using MWNTs as a modifier. *Macromole. Symposia.* **2007**, *249–250(1)*, 610–617.

64. Wang, Z.; and Ma, J.; The role of nonsolvent in-diffusion velocity in determining polymeric membrane morphology. *Desalinat.* **2012**, *286(0)*, 69–79.
65. Vilatela, J. J.; Khare, R.; and Windle, A. H.; The hierarchical structure and properties of multifunctional carbon nanotube fibre composites.*Carbon.* **2012**, *50(3)*, 1227–1234.
66. Benavides, R. E.; Jana, S. C.; and Reneker, D. H.; Nanofibers from scalable gas jet process. *ACS Macro Lett.* **2012**, *1(8)*, 1032–1036.
67. Gupta, V. B.; and Kothari, V. K.; Manufactured Fiber Technology. Springer; **1997**, 661 p.
68. Wang, T.; and Kumar, S.; Electrospinning of polyacrylonitrile nanofibers. *J. Appl. Polym. Sci.* **2006**, *102(2)*, 1023–1029.
69. Song, K.; et al. Lubrication of poly (vinyl alcohol) chain orientation by carbon nanochips in composite tapes. *J.Appl. Polym. Sci.* **2013**, *127(4)*, 2977-2982.
70. Theodorou, D. N.; Molecular simulations of sorption and diffusion in amorphous polymers. *PlasticsEng.-New York.* **1996**, *32*, 67–142.
71. Müller-Plathe, F.; Permeation of polymers—a computational approach. *Acta Polymer.* **1994**, *45(4)*, 259–293.
72. Liu, Y. J.; and Chen, X. L.; Evaluations of the effective material properties of carbon nanotube-based composites using a nanoscale representative volume element. *Mech. Mater.* **2003**, *35(1)*, 69–81.
73. Gusev, A. A.; and Suter,U.W.; Dynamics of small molecules in dense polymers subject to thermal motion. *J.Chem. Phys.***1993**, *99*, 2228.
74. Elliott, J. A.; Novel approaches to multiscale modelling in materials science. *Int. Mater. Rev.* **2011**, *56(4)*, 207–225.
75. Greenfield, M. L.; and Theodorou, D. N.; Molecular modeling of methane diffusion in glassy atactic polypropylene via multidimensional transition state theory. *Macromole.* **1998**, *31(20)*, 7068–7090.
76. Peng, F.; et al. Hybrid organic-inorganic membrane: solving the tradeoff between permeability and selectivity. *Chem. Mater.* **2005**, *17(26)*, 6790–6796.
77. Duke, M. C.; et al. Exposing the molecular sieving architecture of amorphous silica using positron annihilation spectroscopy. *Adv. Funct. Mater.* **2008**, *18(23)*, 3818–3826.
78. Hedstrom, J. A.; et al. Pore morphologies in disordered nanoporous thin films. Langmuir, **2004**, *20(5)*, 1535–1538.
79. Pujari, P. K.; et al. Study of pore structure in grafted polymer membranes using slow positron beam and small-angle x-ray scattering techniques.*Nuclear Inst. Methods Phys. Res. Section B: Beam Int. Mater. Atoms.* **2007**, *254(2)*, 278–282.
80. Wang, X. Y.; et al. Cavity size distributions in high free volume glassy polymers by molecular simulation.*Polymer.* **2004**, *45(11)*, 3907–3912.
81. Skoulidas, A. I.; and Sholl, D. S.; Self-diffusion and transport diffusion of light gases in metal-organic framework materials assessed using molecular dynamics simulations.*J. Phys. Chem. B.* **2005**, *109(33)*, 15760–15768.
82. Wang, X. Y.; et al. A molecular simulation study of cavity size distributions and diffusion in para and meta isomers.*Polymer.* **2005**, *46(21)*, 9155–9161.
83. Zhou, J.; et al. Molecular dynamics simulation of diffusion of gases in pure and silica-filled poly (1-trimethylsilyl-1-propyne)[ptmsp].*Polymer.* **2006**, *47(14)*, 5206–5212.

84. Scholes, C. A.; Kentish, S. E.; and Stevens, G. W.; Carbon dioxide separation through polymeric membrane systems for flue gas applications.*Recent Patents Chem. Eng.* **2008,** *1(1),* 52–66.

85. Wijmans, J. G.; and Baker, R. W.; The solution-diffusion model: a unified approach to membrane permeation. *Mater. Sci. Membranes Gas Vapor Separat.* **2006,** 159–190.

86. Wijmans, J. G.; and Baker, R. W.; The solution-diffusion model: a review.*J. Membrane Sci.* **1995,** *107(1),* 1–21.

87. Way, J. D.; and Roberts, D. L.; Hollow fiber inorganic membranes for gas separations. *Separat. Sci.Technol.***1992,** *27(1),* 29–41.

88. Rao, M. B.; and Sircar, S.; Performance and pore characterization of nanoporous carbon membranes for gas separation. *J. Membrane Sci.* **1996,** *110(1),* 109–118.

89. Merkel, T. C.; et al. effect of nanoparticles on gas sorption and transport in poly (1-trimethylsilyl-1-propyne). *Macromole.* **2003,** *36(18),* 6844–6855.

90. Mulder, M.; Basic Principles of Membrane Technology Second Edition. Kluwer Academic Publication; **1996,** 564 p.

91. Wang, K.; Suda, H.; and Haraya, K.; Permeation time lag and the concentration dependence of the diffusion coefficient of co2 in a carbon molecular sieve membrane. *Ind. Eng. Chem. Res.* **2001,** *40(13),* 2942–2946.

92. Webb, P. A.; and Orr, C.; Analytical Methods in Fine Particle Technology. Micromeritics Norcross, GA;**1997,** *55,*301 p.

93. Pinnau, I.; et al. Long-term permeation properties of poly (1-trimethylsilyl-1-propyne) membranes in hydrocarbon—vapor environment. *J. Polym. Sci. Part B: Polym. Phys.***1997,** *35(10),* 1483–1490.

94. Jean, Y. C.; Characterizing free volumes and holes in polymers by positron annihilation spectroscopy. *Positron Spectroscopy of Solids.* **1993,**1.

95. Hagiwara, K.; et al. Studies on the free volume and the volume expansion behavior of amorphous polymers. *Radiat. Phys. Chem.***2000,** *58(5),* 525–530.

96. Sugden, S.; Molecular volumes at absolute zero. part ii. zero volumes and chemical composition. *J. Chem. Soc. (Resumed).* **1927,** 1786–1798.

97. Dlubek, G.; et al. Positron Annihilation: A Unique Method for Studying Polymers. in Macromolecular Symposia. Wiley Online Library; **2004.**

98. Golemme, G.; et al. NMR study of free volume in amorphous perfluorinated polymers: comparsion with other methods. *Polym.***2003,** *44(17),* 5039–5045.

99. Victor, J. G.; and Torkelson, J. M.; On measuring the distribution of local free volume in glassy polymers by photochromic and fluorescence techniques. *Macromole.* 1987,*20(9),* 2241–2250.

100. Royal, J. S.; and Torkelson, J. M.; Photochromic and fluorescent probe studies in glassy polymer matrices. Macromolecules, **1992,** *25(18),* 4792–4796.

101. Yampolskii, Y. P.; et al. Study of high permeability polymers by means of the spin probe technique. *Polym.* **1999,** *40(7),* 1745–1752.

102. Kobayashi, Y.; et al. Evaluation of polymer free volume by positron annihilation and gas diffusivity measurements. *Polym.* **1994,** *35(5),* 925–928.

103. Huxtable, S.T.; et al. Interfacial heat flow in carbon nanotube suspensions.*Nature Mater.* **2003,** *2(11),* 731–734.

104. Allen, M. P.; and Tildesley, D. J.; Computer Simulation of Liquids. Oxford University Press; **1989.**

105. Frenkel, D.; Smit, B.; and Ratner, M. A.; Understanding molecular simulation: from algorithms to applications. *Phys. Today.* **1997,** *50,* 66.
106. Rapaport, D. C.; The art of Molecular Dynamics Simulation. Cambridge University Press; **2004,** 549.
107. Leach, A. R.; and Schomburg, D.; Molecular Modelling: Principles and Applications. Longman London; **1996.**
108. Martyna, G. J.; et al. Explicit reversible integrators for extended systems dynamics. *Mole. Phys.* **1996,** *87(5),* 1117–1157.
109. Tuckerman, M.; Berne, B. J.; and Martyna, G. J.; Reversible multiple time scale molecular dynamics. *J.Chem. Phys.* **1992,** *97(3),* 1990.
110. Harmandaris, V. A.; et al. Crossover from the rouse to the entangled polymer melt regime: signals from long, detailed atomistic molecular dynamics simulations, supported by rheological experiments. *Macromole.* **2003,** *36(4),* 1376–1387.
111. Firouzi, M.; Tsotsis, T. T.; and Sahimi, M.; Nonequilibrium molecular dynamics simulations of transport and separation of supercritical fluid mixtures in nanoporous membranes. I. Results for a single carbon nanopore. *J. Chem. Phys.* **2003,** *119,* 6810.
112. Shroll, R. M.; and Smith, D. E.; Molecular dynamics simulations in the grand canonical ensemble: application to clay mineral swelling. *J.Chem. Phys.* **1999,** *111,* 9025.
113. Firouzi, M.; et al. Molecular dynamics simulations of transport and separation of carbon dioxide–alkane mixtures in carbon nanopores. *J.Chem. Phys .* **2004,** *120,* 8172.
114. Heffelfinger, G. S.; and van Swol, F.; Diffusion in Lennard-Jones fluids using dual control volume grand canonical molecular dynamics simulation (DCV-GCMD). *J. Chem. Phys.* **1994,** *100,* 7548.
115. Pant, P. K.; and Boyd, R. H.; Simulation of diffusion of small-molecule penetrants in polymers. *Macromole.* **1992,** *25(1),* 494–495.
116. Allen, M. P.; and Tildesley, D. J.; Computer Simulation of Liquids. Oxford University Press; **1989,** 385.
117. Cummings, P. T.; and Evans, D. J.; Nonequilibrium molecular dynamics approaches to transport properties and non-newtonian fluid rheology. *Ind. Eng. Chem. Res.* **1992,** *31(5),* 1237–1252.
118. MacElroy, J.; Nonequilibrium molecular dynamics simulation of diffusion and flow in thin microporous membranes. *J.Chem. Phys.* **1994,** *101,* 5274.
119. Furukawa, S.; and Nitta, T.; Non-equilibrium molecular dynamics simulation studies on gas permeation across carbon membranes with different pore shape composed of micro-graphite crystallites. *J. Membrane Sci.* **2000,** *178(1),* 107–119.
120. Düren, T.; Keil, F. J.; and Seaton, N. A.; Composition dependent transport diffusion coefficients of ch< sub> 4</sub>/cf< sub> 4</sub> mixtures in carbon nanotubes by non-equilibrium molecular dynamics simulations. *Chem.Eng. Sci.* **2002,** *57(8),* 1343–1354.
121. Fried, J. R.; molecular simulation of gas and vapour transport in highly permeable polymers. *Mater. Sci. Membranes Gas Vapour Separat.* **2006,** 95–136.
122. El Sheikh, A.; Ajeeli, A.; and Abu-Taieh,E.; Simulation and Modeling: Current Technologies and Applications. IGI Publishing; **2007.**
123. McDonald, I.; NpT-ensemble monte carlo calculations for binary liquid mixtures. *Mole. Phys.* **2002,** *100(1),* 95–105.

124. Vacatello, M.; et al. A computer model of molecular arrangement in a n-paraffinic liquid. *J.Chem Phys.* **1980,** *73(1),* 548–552.
125. Furukawa, S.-I.; and Nitta, T.; Non-equilibrium molecular dynamics simulation studies on gas permeation across carbon membranes with different pore shape composed of micro-graphite crystallites. *J. Membrane Sci.***2000,** *178(1),* 107–119.

CHAPTER 5

EFFECTS OF POLYPROPYLENE FIBERS AND NANOPARTICLES ON MICROSTRUCTURE OF FIBER REINFORCED CONCRETE

M. MEHDIPOUR

CONTENTS

5.1 INTRODUCTION

Concrete is widely used in structural engineering due to its high compressive strength, low-cost and abundant raw material. But common concrete has some shortcomings, for example, shrinkage and cracking, low tensile and flexural strength, poor toughness, high brittleness, low shock resistance and so on, that restrict its applications. It is well that documented some mechanical properties of concrete such as compressive and flexural strength and abrasion resistance can be significantly improved by addition fibers like polypropylene (PP), glass, carbon, and nanomaterials such as Al_2O_3, TiO_2, SiO_2, and carbon nanotube (CNT). It is believed that application of these nanoparticles in cement matrix up to 3 wt. % could accelerate formation crystalline silicate regions in reaction between nano-SiO_2 and cement matrix (C-S-H) as a result of increased crystalline $Ca(OH)_2$ amount at the early age of hydration and hence increase compressive strength of concrete [1–4]. Vital role of PP fiber creates some connecting bridge between cracks in the concrete microstructure. These cracks produce for some reasons such as alkaline hydrated reaction of cement in concrete paste, thermal contraction, drying shrinkage, and autogenously shrinkage. Fiber-reinforced concrete (FRC) has been successfully used in construction with its excellent flexural-tensile strength, resistance to splitting, impact resistance and excellent permeability. It is also an effective way to increase toughness, shock resistance, and resistance to plastic shrinkage cracking of the mortar [2, 3]. The uniform dispersion of the fibers and other additives is an important viewand without this property, addition of fiber and any other additives have inverse effect on mechanical properties of concrete [5, 6].

The performance of PP fibers as reinforcement in cement-based materials for different fiber length and proportion has been investigated by Bagherzadeh et al [3, 4]. It was concluded that PP fibers, in spite of their hydrophobic nature and weak wetting by cement paste, show better bonding than other fibers. Also, it was concluded that with an increase in fiber length or decrease in fiber diameter, crack width decreases significantly. It is also reported [6, 9] that adding PP fibers in concrete materials decrease internal pressure of concrete and can enhance the fire resistance of concrete structures, since it can be helpful to avoid explosive spalling.

Nanoscale materials due to their special property such as high surface to volume ratios have high ability to improve the physical and mechanical properties of cementitious material [3, 4]. Nano-SiO_2 with range diameter less than 100 nm is one of the most common nanoparticles in concrete industry. Addition of nano-SiO_2 to mixture of concrete cause activation of SiO_2 which can react with $Ca(OH)_2$ of cement rapidly and produce C-S-H that decrease porosity of cement matrix. Formation of C-S-H in the cement matrix improves mechanical properties of concrete [13]. Also nano-SiO_2 creates a net in interfacial transition zone (ITZ), a small region beside the particles of coarse aggregate and cement, can make it more homogeneous which can improve concrete microstructure and increase its workability[5,14]. Nanoparticles dispersed in the cement paste uniformly, could accelerate cement hydration due to their high-activity [8]. However, when nanoparticles cannot be well dispersed, their aggregation will create weak zones in the cement matrix and consequently, the homogeneous hydrated microstructure in ITZ could not be formed and low mechanical properties will be expected [2–8]. Only few in-depth studies related to nanomodifications of cement based materials and their enhanced performance on compressive and tensile strength properties are reported in literature. However, this field requires bottom-up approach to tailor, modify, replace, or include nanomaterials in the already available matrices such as cement or cement-based materials. Another nanoparticle material which is believed that can improve the mechanical properties of cementitious materials is nano-TiO_2. Addition of the nano-TiO_2 with its tetragonal crystalline structure can improve some mechanical properties such as abrasion resistance of concrete. Due to its ability to produce OH and O_2 radicals with photo catalytic reactions, it could be useful to achieve self-cleaning effect in surface of cement-based composites [7]. It is also reported that flexural strength of the concrete with 1 per cent weight of nano-TiO_2 is much better than the concrete with only pp fibers or nano-SiO_2 [12].

There are well-documented reports in the area of adding fibers and nanoparticles in concrete; however, there is still a lack of studies on the best volume fractions of fibers and particles in the concrete mixture. Also there is a lack of study in investigation of simultaneously effect of fibers and nanoparticles in mixture design of cementitious materials. This study presents comprehensive experimental data regarding the effects of adding PP fibers with different fiber lengths and diameter, and different volume fractions, along with different nanoparticle types and portions on the physical and mechanical properties of FRC.

5.2   MATERIALS AND METHODS

5.2.1   MATERIALS

PP fibers used in this study were provided by the Shimifaraiand Company. Specifications of the fibers employed in this study are presented in Table (5.1). Ordinary Portland cement conforming to ISO 12269 was used for the concrete mixtures. River sand with a specific gravity of 2.65 (g/cm^3) and fineness modulus of 2.64 was used as the fine aggregate. A commercially water reduction agent (SP), Glenium—110 p (because w/c in mixture design should be stayed constant)—was purchased from BASF company. Naphthalene- based super plasticizer from LanYa Concrete Admixtures Company was purchased too. Nanoparticles including TiO_2 and SiO_2 were acquired from the Degussa Co. (Germany). Specifications of the nanoparticles employed in this study are presented in Table (5.2).

TABLE 5.1   Properties of polypropylene fiber and nano-SiO_2 and nano-TiO_2 particles

	Density (g/cm^3)	Tensile strength (MPa)	Elastic modulus (GPa)	Length (mm)	denier
PP fiber	0.9	350	3.4–3.6	12	5
	Diameter (nm)	Specific surface area (m²/g)	Density (g/cm^3)	Purity (%)	
SiO_2	12 3	200 5	0.6	99.5	
TiO_2	15 5	50 5	0.75	99.4	

5.2.2   MIXING AND CURING

Trial mixtures were prepared to obtain target strength of 40 MPa at 28 days, along with a good workability. The final mixture design used in this study was composed of 350 kg/m^3 cement, 184 kg/m^3 water, 0.50 water/cement percentage (W/C), and 1.35 per cent naphthalene- based super plasticizer. The total dosage of fibers was maintained at 0–3 per cent, primarily from the point of view of providing good workability and no balling of fibers during mixing.

The sand, cement, and other additives were first mixed dry in a pan mixer with a capacity of 100 kg for a period of 2 min. The naphthalene-based super plasticizer and nanoparticles then mixed thoroughly with the mixing water and added to the mixer in order to avoid the agglomeration of nanoparticles. First, fibers were dispersed by hand and then were mixed in the mixture for a total time of 4 min to achieve a uniform distribution throughout the concrete. Fresh concrete was cast in steel molds and compacted on a vibrating table. For the curing function, the specimens were kept covered in their molds for 24 h. The specimens had less bleeding than the control concrete specimen. After demolding, concrete specimens were placed in $20 \pm 2°C$ water for 28 days for curing. They were removed from water and placed in the laboratory environment for 2 h before carry out the tests. All tests were performed according to relevant standards.

5.2.3   PHYSICAL AND MECHANICAL TESTS

5.2.3.1   COMPRESSIVE STRENGTH

A universal testing machine with a capacity of 100 tones was used for testing the compressive strengths of all $100 \times 100 \times 100$ mm cube Specimens. These specimens at 7-day and 28-days from casting were tested at a loading rate of 14 $N/mm^2/min$ according to BS 1881–116 standard. The compressive strength was interpreted as the stress generated from the result of compression load per area of specimen surface. The results for each specimen are based on an average value of three replicate Specimens.

5.2.3.2   FLEXURAL STRENGTH

Flexural strength at 28-days of curing test was conducted according to the requirements of ASTM C 293 using three $100 \times 100 \times 500$ mm beams under mid-point loading on a simply supported span of 300 mm. According to ASTM standards, the results of the flexural strength test are interpreted by calculating flexural stress as follows:

$$R = \frac{PL}{bd^2} \tag{5.1}$$

where R is the flexural strength (modulus of rupture), P is the maximum indicated load, L is the span length, b is the width of the Specimen, and d is the depth of the Specimen.

5.2.3.3   WATER ABSORPTION

Water penetration at 28-day of curing tests of two $100 \times 100 \times 100$ mm concrete mixture ‹optimum and plain sample was conducted according to the requirements of ASTM C1585 ,in due to difference between wet and dry sample of concrete.

5.2.4   DESIGN OF EXPERIMENTS

The use of statistical methods has helped the rapid development of nanotechnology in terms of data collection, hypothesis testing, and quality control. Response Surface Methodology (RSM) is an empirical modeling technique used to evaluate the relation between a set of controllable experimental factors and observed results [11]. At first stage of this study, concrete mixture samples designed by D-optimal method. D-Optimal designs are based on a computer-aided exchange procedure which creates the optimal set of experiments. D × 7 (statistic's software , that selections optimized samples and conditions due to design of experiments) selected 11 mixture designs, and two cubic specimens from each mixture design were made to measure compressive strength both at 7 days and 28 days after casting.

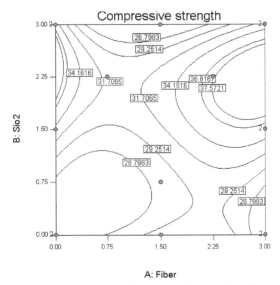

FIGURE 5.1 RSM plots for compare effects of polypropylene fiber and nano-SiO$_2$ on 7-day compression strength of samples (MPa).

In second stage, nano-TiO$_2$ added to 3 mixture samples simultaneously with nano-SiO$_2$and Polypropylene fiber. Eight prepared samples were tested to evaluate the 28-day compressive and 28-day flexural strengths.

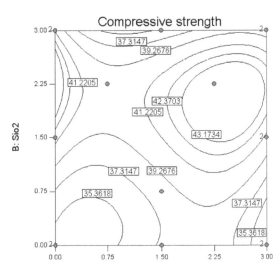

FIGURE 5.2 RSM plots for compared effects of Polypropylene fiber and nano-SiO$_2$ on 28-day compression strength of samples (MPa).

5.2.5 SCANNING ELECTRON MICROSCOPE

The concrete specimen cured after 90 days was broken into small pieces. Those pieces with dimensions around 7 mm by 7 mm by 4 mm and containing aggregates and tri-calcium silicate (3CaO·SiO2) were selected as samples for the SEM experiment because expected that this products showed optimum results. These samples were washed using 1 per cent HCL for 20 s, and then dried at 60–70°C. Consequently, the absolute ethyl alcohol was employed to stop hydration of concrete samples which were then coated with golden conductive films in the vacuum coating machine (DMX – 220A) using the vaporization method.

The SEM (JSM – 840) was employed to observe the microstructures of samples. In this experiment, the interface between the aggregate and the cement paste was positioned at the center of the SEM in order to observe the interfacial transition zone. Figure (5.3) shows the microstructures of the plain and nanoparticles and fibers reinforced concretes (NF) and its microstructure's of the aggregate-cement interfacial transition zone with magnification of 5,000x.

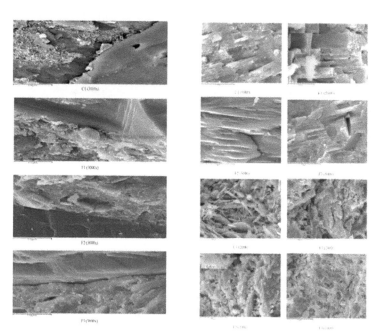

FIGURE 5.3   Microstructures of the plain and NF reinforced concretes with magnification of 5,000x.

5.3 RESULTS AND DISCUSSIONS

5.3.1 MECHANICAL AND PHYSICAL PROPERTIES

Concrete mixture proportions and the measured mechanical and physical properties including compressive strength, flexural strength and water penetration of both the plain concrete and NF (nanoparticles and fibers reinforced concretes) mixtures are presented in Tables (5.2–5.5).

TABLE 5.2 Concrete mixture proportions used in first stage of this study

Sample code	Fiber (F) or nano-Sio$_2$ (NS) content
P1	1.5% F[a], 0.75% NS[b]
P2	3% F, 3% NS
P3	3 % F, 1.5% NS
P4	2.25% F, 2.25% NS
P5	0% F , 0 % NS
P6	0% F, 3% NS
P7	1.5% F, 0%NS
P8	0.75 % F, 2.25% NS
P9	0% F, 1.5% NS
P10	1.5% F, 3% NS
P11	3% F, 0%NS

[a]F refers to the weight per cent of cement for PP fiber

[b]NS refers to the weight per cent of cement for nano-SiO$_2$

TABLE 5.3   Result of concrete mixtures used in first stage of this study

Specimens code	P1	P2	P3	P4	P5	P6	P7	P8	P9	P10	P11
Compression Strength (MPa) (7 days)	25	27	33	36	36	27	34	31	31	34	23
Compression Strength (MPa) (28 days)	38	38	41	44	33	41	38	40	42	36	30

TABLE 5.4   Concrete mixture proportions used in second stage of this study

Sample code	Weight percentage of cement in mixture for Fiber (F), nano-Sio$_2$ (NS), and nano-TiO$_2$ (NT) contents in
a	2.25% F, 0.75% NT, 0% NS
b	2.25% F, 0.75% NT, 0.75% NS
c	2.25% F , 2.25% NT, 2.25% NS
d-1	2.25% F, 2.25% NS
d-2	2.25% F, 2.25% NS
e-1	0% F, 0% NS
e-2	0% F, 0% NS
f	2.25% F, 0% NS

TABLE 5.5   Results of compression and flexural strength and water absorption for second stage specimens

Specimens code	a	b	c	d-1	d-2	e-1	e-2	f
Compression Strength (MPa) (28 days)	28.8	40.9	38	45.4	45.7	31.6	31.6	32.4
Flexural Strength (MPa) (28 days)	–	–	–	10.5	12	8.5	11.1	10.7
Water absorption	Plain (2.4%)				Opti-mum (1.4%)			

5.3.1.1   COMPRESSIVE AND FLEXURAL STRENGTHS

Result showed that fibers and nanoparticles have effectively improved concrete's compressive and flexural strengths, (see Tables 5.3–5.5). For example, in P4 mixture that contains 2.25 per cent of PP fiber and 2.25 per cent of nano-SiO_2, flexural strength is 13 per cent higher, and compressive strength is 25per cent higher than those of plain sample (P5 sample).

It also shows that both compressive and flexural strengths increase to its maximum amount for P4 sample; a good relation about these two factors can be confirmed. This sample also selected as the optimum sample at both age of curing. It also showed that both compressive and flexural strengths decrease with increase in fibers or nanoparticles content. For example, P11 mixture with more fiber content than others even has lower strengths than p5 mixture. It is due to the random contribution of fibers and lack of nanoparticles' good dispersion.

By adding the nano-$SiO_2$to (b) and (c) mixtures, compressive strength is increased, significantly. In other words, nano-SiO_2 is more effective than nano-$TiO_2$to improve the compressive strength. The reason may be attributed to the fact that the specific surface of nano-SiO_2 is much larger than nano-TiO_2, so nano-TiO_2 is more difficult to uniformly disperse than nano-SiO_2 in concrete microstructure. Results of 28-day compressive strength is more than 7-day for all specimens as it was expected. Also by adding PP fibers in the mixture, flexural strength is improved significantly for all specimens. It seems that increase in flexural strength of concrete due to adding PP fiber is more than increase in compressive strength of samples. For p6 and p9 samples that include only nano-SiO_2, both 7-day and 28-day compressive strengths are more than those for p7 and p11 samples that were contained only PP fibers. This shows that nano-SiO_2 is more effective in compressive strength improvement than PP fiber.

RSM plots (Figures 5.1 and 5.2) confirm the results of this test in first stage, since the values were statistically analyzed by the ANOVA technique (variances analyses tables due to design of experiments' method) at the 95 per cent confidence level. The coefficient of determination ($R2$) calculated in 7-day and 28-day compressive strength, was 0.9620 and 0.9288, that showed this model is acceptable.

5.3.1.2 WATER PENETRATION

Water penetration result (Table 5.5) shows that the water penetration depth of concrete has significantly reduced after adding nano-siO_2 (2.25% of PP fiber and 2.25% of nano-SiO_2). Water penetration for P4 sample (1.4%) is much less than that of plain sample (2.4%). Therefore, by adding fibers and nanoparticles to concrete samples, water penetration was decreased. The reason for this could be due to decreasing in porosity and inhibiting the formation of micro cracks in concrete microstructure which can consequently enhance durability of concrete.

5.3.2 SEM RESULTS AND ANALYSIS ABOUT MICROSTRUCTURE OF CONCRETE

The primary phase compositions of hydrated concrete include the amorphous and poorly crystalline calcium silicate hydrate (C–S–H), the layered calcium hydroxide (CH, or $Ca(OH)_2$), and the calcium aluminate/aluminoferrite hydrate phases (AFm and AFt-type phases) [13]. C–S–H is the primary bind phase in the hardened concrete. CH is also an important component occupying about 20–25 per cent of hydrate products by volume [14]. Figure (5.3) shows the microstructures of the plain and optimum modified concrete samples. It seems that C1 mixture (plain concrete) contains many crystalline with the shape like fiber, snow flower, or needle, which interweave together among C–S–H. However, the CH crystalline presents in a layered structure with high orientation. Meanwhile, many voids are noticeable among these crystalline regions. In contrast, optimum sample contains many gel components and have much less voids than C1 sample. Its crystalline size is obviously smaller than that of C1. The size of the needle-shaped crystallines decreases from 8 to 10 μm (C1 mixture) to 2–4 μm (F2 mixture). It even seems hard to identify the interfaces between different crystalline layers. Especially F2 mixture contains very few voids among crystalline, and its gels look very integral and continuous. Figure (5.3) also shows the microstructures of the aggregate-cement interfacial transition zone for the plain and NF reinforced concretes with a magnifying ratio of 3,000. It shows that C1 contains many C–S–H gel and hexagon sill CH crystalline with relatively large particle sizes, and it has noticeable voids and micro-cracking at the interfacial transition zone. However, the microstructures at the interfacial transition zone of the optimum mixtures especially F2 look much denser with more uniform particle sizes than that of C1, the voids at its interfacial transition zone are very minimal (F2 mixture in Figure 5.3).

5.3.3 MECHANISM OF FIBER AND NANOPARTICLES MODIFICATION

It has expected that fiber and nanoparticles performsa kind of network, that make a bridge between cracks that cause dense microstructure and toughens the reinforce concrete, thus fiber and nanoparticles significantly improves the toughness of cementation matrices [2–3–11–12].

It seems that fibers and nanoparticles network's formation is restricted the growth of CH crystalline and reduced the micro voids, and therefore decreases the size and orientation of CH. As a result, the microstructures and the aggregate-cement interfacial transition zone of NF are much denser with less micro-cracking than that of plain concrete, and the size and orientation of CH have significantly reduced. Therefore, NF's compressive and flexural strengths values, have significantly increased, and the water penetration have obviously decreased. Furthermore, fibers and nanoparticles create the networking effect to hold concrete aggregates together and also prevent crack propagation. Fibers would also perform the bridging crack function to take part of the internal tensile stress and thus resist the crack propagation after appearing micro-cracks. In addition, fiber reduces the shrinkage contractions and thus reduces the interfacial relative slides between concrete and other substrate like hardened concrete, which contribute to part of the improved interfacial bonding strength discussed previously. However, when fibers and nanoparticles contents are relatively up to (3%), fibers and nanoparticles distribution in concrete mixture becomes uniform weekly due to the reduced workability.

As a result, concrete density decreases with increasing fibers and nanoparticles contents and the aggregated fibers and nanoparticles may form weak points and induce relatively large voids in the concrete structure. Also, the material strengths and water penetration resistance decrease with increasing fiber and nanoparticles content. However, other research showed that additives like silica fume may improve the dispersion of fiber and nanoparticles in concrete, and thus help improve the performance of concrete samples.

5.4 CONCLUSION

This research endeavors to study how fiber and nanoparticles modify the physical and mechanical properties of concrete. The primary findings are summarized as follows:

1. Simultaneous application of the fiber and nanoparticles can significantly alter the microstructure of concrete as it reduces the amount, size, and orientation of CH crystals and micro voids, and it reduces the voids and micro-cracking at the interfacial transition zone between aggregate and cement;

2. Simultaneous application of the fiber and nanoparticles effectively improves the engineering properties of concrete including material strengths, and water penetration resistance, and fiber specifically reduces the shrinkage contraction of concrete and increases the flexural strength.

3. Simultaneous application of the fiber and nanoparticles forms a network to restrict the growth of CH crystallines and thus condenses the microstructure of concrete and improves physical and mechanical properties of reinforcement concrete.

4. Increase the nanoparticles and pp fiber content over to 3 per cent weight in mixture, showed inverse influence to mechanical properties of concrete. Also it is concluded that nano-SiO_2 is more effective than nano-TiO_2 and PP fiber in improving physical and mechanical properties of concrete. The results indicated that simultaneously adding of nano-particles and PP fibers can significantly improve the physical and mechanical properties of concrete.

KEYWORDS

- Concrete
- Fiber-reinforced concrete
- ITZ
- Nanoparticle
- Physical and mechanical properties

REFERENCES

1. Mazaheripour, H.; Ghanbarpour, S.; "The effect of pp fibers on the properties of fresh and hardended lightweight self-compacting concrete" *Construct. Build. Mater.* **2011**, *25(1)*, 351–358.
2. Raki, L.; Beaudion, J.; Alizadeh, R.; Maker, J.; Sate, T.; "Cement and concrete nano science and nano technology". *Materials.* **2010**, *3*,918–942.
3. Bagherzadeh, R.; Sadeghi, A. H.; Latifi, M.; "Utilizing polypropylene fibers to improve physical and mechanical properties of concrete". *Textile Res. J.* **2012**, *82(1)*,88–96.
4. Bagherzadeh, R.; Pakravan, H. R.;Sadeghi, A. H.;Latifi, M.; Merati, A. A.; "An investigation on adding polypropylene fibers to reinforce lightweight cement composites (LWC)". *J. Eng. Fibers Fabrics.* **2012**, *7(4)*, 13–21.
5. Sun, Z.; and Xu, Q.; *Mater. Sci. Eng. A.* **2009**, *527(1–2)*, 198–204.
6. Behnood, A.; and Ghandehari, M.; "Comparison of compressive and splitting tensile strength of highstrength concrete with and without pp fibers heated to high temperature". *Fire Saf. J.* **2009**, *44(8)*, 1015–1022.
7. Diamanti, M. V; Ormellese, M.; and Pedeforri, M. P.; "Characterization of photo catalytic and super hydrophilic properties of mortars containing TiO_2". *Cem. Concr. Res.* **2008**, *38*, 1349–1353.
8. Li Hui, Z-M Hua; Ou Jing-Ping; "Abrasion resistance of concrete containing nanoparticles", *Wear*, **2006,** *260(11–12)*, 1262–1266.
9. Kodur, V. K. R; Cheng, F-P; Wang, T-C; and Sultan, M-A.; "Effects of strength and fiber reinforcement on the fire resistance of high strength concrete columns". *J. Struct. Eng.* **2003**, *129(2)*, 253–259.
10. Li Hui, Yuon-Jie; and Ou Jing-Ping; "Microstructure of cement mortar with nanoparticles".*Compos. Part B Eng.* **2004**, *35*, 185–189.
11. Jye–Chyi Lu, Kaibo Wang; Shuen–Lin Jeng; "A review of statistical methods for quality improvement and control in nanotechnology". *J. Qual. Technol.* **2009**, *41(2)*.
12. Li Hui, Z-m Hua; Ou Jing-Ping; "Flexural fatigue performance of concrete containing nano-particles for pavement". *Int. J. Fatigue.* **2007**, *29*, 1292–1301.
13. Girao, A. V.; Richardson, I. G.; Porteneuve, C. B.; Brydson, R. M. D.; "Composition, morphology and nanostructure of C–S–H in white Portland cement pastes hydrated at 55°C". *Cem. Concr. Res.* **2007**, *37*, 1571–1582.
14. Richardson, I. G.; "Tobermorite/jennite- and tobermorite/calcium hydroxide-based models for the structure of C–S–H: applicability to hardened pastes of tricalcium silicate, β-dicalcium silicate, Portland cement, and blends of Portland cement with blast-furnace slag, metakaolin, or silica fume". *Cem. Concr. Res.* 2004, *34*, 1733–1777.

CHAPTER 6

IMPROVING OF THE FUNCTIONALITY OF GLYCININ BY LIMITED PAPAIN HYDROLYSIS

A. V. POLYAKOV, A. N. DANILENKO, I. L. ZHURAVLEVA,
I. G. PLASHCHINA, S. V. RUDAKOV, A. S. RUDAKOVA, and
A. D. SHUTOV

CONTENTS

6.1 INTRODUCTION

Glycinin is a storage soybean protein belonging to the legumin family (11S globulin fraction). Legumins have high nutritional value, however, due to the peculiarities of molecular structure (high stiffness, low solubility near the isoelectric point) and, consequently, not enough high-functional properties, their use is limited.

One of the most soft and promising ways of improving the functional properties of proteins is limited proteolysis. It is caused by the presence in a substrate molecule bonds (sites) are sensitive to attack with proteolytic enzymes. Limited proteolysis slows down in descending order with the formation of relatively stable high-molecular-weight product. The reorganization of protein structure by limited proteolysis leads to a change in its physico-chemical and functional properties.

The purpose of this study was to establish the interrelationships between the change in molecular parameters of glycinin as a result of limited papain hydrolysis and a change of its there most ability, surface behavior at the air–water interface (surface activity, the dynamics of formation, and dylatation properties of adsorption layers).

6.1.1 MATERIALS AND METHODS

Samples of intact glycinin-11S globulin fraction of soybean Glycine max (glycinin) and its modificate, obtained by limited proteolysis with papain (glycinin-P), were obtained according (1). The protein was stored in a saturated solution of ammonium sulfate at 5°C.

6.1.2 SOLUTION PREPARATION

The samples of protein solutions were prepared by equilibrium dialysis against 0.05 M phosphate buffer pH 7.6 and 0.5 M NaCl. Solutions before the measurements were centrifuged at 20,000 g for 1 h and filtered through a cellulose membrane Millipore filters with a pore diameter of 0.22μm. The protein concentration in solutions was determined by microbiuret method (2). To prevent microbial contamination 0.02 per cent NaN_3 was added, to reduce the disulfide exchange—0.1 per cent mercaptoethanol.

6.1.3   LASER LIGHT SCATTERING

Determination of molecular parameters (molecular weight, second virial coefficient, hydrodynamic size, ξ-potential) was performed by static and dynamic laser light scattering using equipment Zeta Sizer Nano (ZEN 3600) (Malvern Instruments Ltd., UK), equipped with a 4 MW He–Ne laser (λ_0 = 632,8 nm).

The measurements were performed at 25 C and fixed scattering angle of 173 .

A sample solution was filtered through a Millipore filter of 22 μm into a 1 cm measuring cuvette, which was preliminary washed with ~1mL of the filtrate. The time of temperature equilibration of the cuvette inside the instrument was 10 min. The final hydrodynamic radius distribution of the sample was found by averaging 10 measurements, each of which was a result of 10–16 scans.

6.1.4   SMALL-ANGLE X-RAY SCATTERING

Intensity of small-angle X-ray scattering of glycinin and glycinin-P solutions was measured at the ambient temperature ~25 C with the diffractometer described elsewhere [2-3] in the range of S values from 0.15 to 3 nm^{-1}, where $S = (4\pi\sin\theta)/\lambda$, θ–half of a scattering angle, λ–CuKα wavelength (0.1542 nm). The buffer used was 0.05M phosphate buffer saline (pH7.6) containing 0.5M NaCl, 0.02 per cent NaN$_3$ and 0.1 per cent 2-merkapto-ethanol. Protein concentration was ~30 mg/ml. Data processing including smoothing and desmearing was done as in [3]. The protein radius of gyration R_g was determined with the PRIMUS software [4] utilizing intensity $I(S)$ approximation with the Guinier formula [1-5]:

$$I(S) = I(0)\exp\{-(SR_g)^2/3\}$$

6.1.5   DIFFERENTIAL SCANNING MICROCALORIMETRY

Determination of the thermodynamic parameters of thermodenaturation carried out by the method of adiabatic differential scanning microcalorimetry with using of microcalorimeter DASM-4 (BIOPRIBOR, Russia) within the temperature range 10–130°C at a heating rate of 2 K min−1 and an excess pressure of 0.25 MPa. The primary data processing and conversion of the partial heat capacity of protein into the excess heat capacity function of the denaturation transition was performed using the standard Wscal software. The baseline in the transition area was obtained by a spline interpolation. The maximum temperature of the excess heat capacity curve was taken as the denaturation temperature, Td. The denaturation enthalpy, ΔH_d^{cal}, kJ/mol, was determined by integration of the excess heat capacity function.

6.1.6 DYNAMIC TENSIOMETRY AND DYLATOMETRY

Investigation of the surface activity, the dynamics of formation, and dylatometric properties of adsorption layers of intact and modified proteins at the air/water interface were carried out on droplet tensiometer Tracker (TRACKER, IT Concept, Longessaine, France). Measurements were made in thermostated cuvette at 25 ± 0.1°C. Two-dimensional complex modulus of elasticity E of the adsorption layer is determined from the measured values of surface tension when the load is applied at the interface, which varies sinusoidally. Measurements of rheological properties produced after the formation of a sufficiently stable adsorption layer within 60–70 000 seconds. Fluctuations in the surface area of the bubble produced with an amplitude of 3 per cent. The series of 3–5 consecutive loads of active and passive cycles at constant amplitude and frequency were carried out. The range of frequencies used was 0.001–0.02 Hz[6-8].

6.1.7 RESULTS AND DISCUSSION

6.1.7.1 GLYCININ AND GLYCININ-P MOLECULAR PARAMETERS

Table 6.1 shows the experimental values of the molecular structure parameters of the intact and modified glycinin (molecular weight, second virial coefficient, effective hydrodynamic radius, radius of gyration, sedimentation coefficient, frictional ratio, and diffusion coefficient, respectively).

TABLE 6.1 Molecular parameters of intact and modified by limited papain hydrolysis glycinin. Conditions: 0.05 M phosphate buffer, pH 7.6, 0.5 M NaCl; Temperature of 25 C

Sample	M_w, kDa	$B_2 \cdot 10^{-4}$ mL·mol/g^2	R_h, nm	R_g, nm	$S\,10^{13}$, c	f/f_0^*	f/f_0^{**}	$D \cdot 10^7$, cm/sec
Glycinin	362 ± 21	3.57 ± 0.21	5.69 ± 0.18	4.22	11.9	1.35	1.43	4.06
Glycinin-P	269 ± 24	1.64 ± 0.18	5.16 ± 0.15	4.15	11.8	1.24	1.26	4.38

Note: f/f_0^*—calculated as the ratio of hydrodynamic radius to the radius of gyration; f/f_0^{**}—calculated on the basis of the equation $f/f_0 = S_{max}/S = 0.00361M^{2/3}/S$ (6.6)

Obviously, during the limited proteolysis some parameters of glycinin reduce: molecular weight, the diffusion coefficient, effective hydrodynamic size, and to a lesser extent—the radius of gyration. The magnitude of frictional relationship suggests lowering the degree of asymmetry of the modified protein molecules compared with intact. It was also established that glycinin-P molecule is characterized by a lower charge density (lower absolute value of ξ-potential: −17.2mV comparatively to −20.3mV of glycinin) and a lower affinity to the solvent, namely, the value of B_2 of glycinin-P is lower than the B_2 of the intact protein molecule (Table 6.1).

Table 6.2 summarizes our obtained glycinin and glycinin-P thermodynamic parameters of denaturation (denaturation temperature, enthalpy, heat capacity increment, the maximum heat capacity, Gibbs free energy, and the parameter of cooperativity, respectively).

From Table 6.2, it follows that glycinin-P thermodynamic parameters of denaturation are lower than that of the intact glycinin, indicating a destabilization of the molecules by limited proteolysis. Denaturation process is less cooperative (increased value of the parameter of cooperativity). Increase the heat capacity increment of glycinin-P compared with glycinin means the growth for the solvent accessible hydrophobic surface. This result correlates with a decrease in the thermodynamic affinity of the solvent glycinin molecules (lower B_2) during proteolysis (Table 6.1).

TABLE 6.2 Thermodynamic parameters of glycinin and glycinin-P denaturation. Conditions: 0.05 M phosphate buffer, pH 7.6, 0.5 M NaCl; Scan rate 2 deg/min, pressure 0.2 MPa, protein concentrations—2 mg/ml

Sample	T_d, K	ΔH_d^{cal}, 10^{-3} kJ/mol	ΔC_p, kJ/ mol·K	Amp, kJ/ mol·K	ΔG_d·10^2 kJ/ mol	$\Delta T = \Delta H_d^{cal}$/ Amp
Glycinin	370,1	12,6	151	1406	13,0	9,0
Glycinin-P	364,9	9,2	173	763	7,5	12,0

It is seen that observed changes in glycinin molecular characteristics as a result of limited proteolysis with papain are favorable for the increase of its surface activity and the rate of formation of the adsorption layers.

6.1.8 SURFACE ACTIVITY AND THE DYNAMICS OF FORMATION OF ADSORPTION LAYERS AT THE AIR–WATER INTERFACE

Dynamic surface tension curves of adsorption layers formed by glycinin and glycinin-P at the air/water solution interface with different protein concentrations in phosphate buffer 0.05 M (pH 7, 6) are shown in Figure 6.1. Shape of the curves is typical for globular proteins. Dependence for both proteins at low concentrations (0.001 and 0.01 mg/ml) are character-ized by an S-shaped form, which can be divided into 3 main sections—an induction period (the initial part of the relatively slow decrease of surface tension), part of a significant reduction in surface tension (post induction period) and the stationary phase, in which the tension reaches a minimum value and subsequently remains practically unchanged. It is known that the presence of the induction period (data are not shown) suggests that the adsorption process is controlled by the diffusion stage of the protein macromolecules from the bulk to the interface. When the concentration of proteins in solution is 0.001 mg/ml, the induction period is 1,400 sec for glycinin and 600 sec for glycinin-P (data is not presented). The decrease of the induction period in the case of glycinin-P is due to the higher rate of diffusion of the molecules caused by the lower molecular weight and a smaller effective hydrodynamic size.

With the increase of the volume concentration of the protein value of the induction period is reduced Eq. (6.7). In our case, at a constant bulk protein concentration of 0.1 mg/ml, the curves $\sigma(\tau)$ show only a postinduction and the final stages of adsorption. At the higher concentrations, the "induction" period is absent, that is, diffusion of macromolecules from the bulk solution to the interface is instantaneous. At the final stage of adsorption, already formed adsorption layer begins to act as a repulsive barrier with respect to the macro-ion, which reaches the surface. This is manifested in a gradual decrease in the rate of $d\sigma/d\tau$. The decrease of surface tension after long periods of time indicates that the adsorbed macromolecule is controlled by diffusion of the active segments of its own redistribution in the adsorption layer. Increasing the rate of glycinin-P adsorption of $d\sigma/d\tau$, as well as a lower value of the quasi-equilibrium values of surface tension of its solutions as compared to glycinin observed throughout the concentration range of proteins. The observed effects correlated with changes in molecular parameters: a lower charge density, higher surface hydrophobicity, and lower conformational stability of glycinin-P comparatively to glycinin. We can assume that glycinin-P is able to generate more perfect adsorption layers.

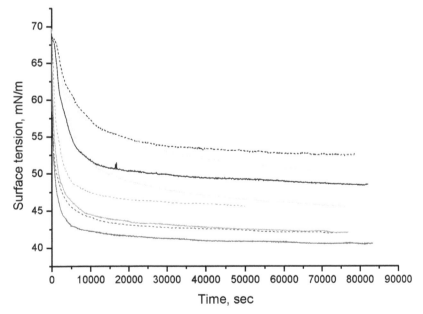

FIGURE 6.1 Dynamic surface tension curves of adsorbed layers of glycinin (dashed lines) and glycinin-p (solid lines) molecules. Protein concentrations: 0.001 mg/ml (black lines), 0.01 mg/ml (green line), 0.1 mg/ml (red line), 1.0 mg/ml (blue line). Solvent—0.05 M phosphate buffer pH 7.6, 0.5 M NaCl, 0.02 per cent NaN_3. Temperature: 25 C.

6.1.9 EFFECT OF PROTEIN CONCENTRATION ON THE EXPANSION JOINT PROPERTIES OF ADSORPTION LAYERS AT THE AIR/SOLUTION INTERFACE

The study of glycinin and glycinin-P rheological properties of adsorption layers was carried out by applying a small longitudinal perturbations ΔA (t) to the surface of an air bubble sinusoidally $\varepsilon\ (t) = \varepsilon_a \bullet \exp\ (i\omega t)$ (where $\varepsilon = \Delta A/A$—relative longitudinal deformation of the layer, ε_a— the amplitude of the strain) at a constant frequency of ω. The deformation causes a change in surface tension-$\Delta\sigma\ (t) = \sigma_a \bullet \exp\ (i\omega t + \varphi)$), where φ—the phase angle Eq. (6.8). In the field of linear viscoelasticity, complex modulus of elasticity can be expressed as follows:

$$\bar{E}(\omega) = \frac{d\bar{\pi}}{d\bar{\varepsilon}} = -\frac{d\bar{\sigma}}{d\bar{\varepsilon}} = E'(\omega) + iE''(\omega)$$

where $E'(\omega)$ and $E''(\omega)$—real (modulus of storage) and imaginary (modulus of lost) parts of which are dependent on the applied frequency ω and define conservative and dissipative nature of the rheology of adsorption layers, respectively.

Figure 6.2 shows the frequency dependence of the applied compression—dilatation strain of the real E' and imaginary E'' components of the quasi equilibrium (formed during the 70,000 sec.) complex viscoelastic modulus of glycinin and glycinin-P adsorption layers at various protein concentrations. In the whole frequency range, both intact and modified forms of glycinin are characterized by a symbatic change in complex module and its real part (E'), and $\varphi > 0$ (data not shown), value of $E' \gg E''$, and E' is weakly dependent on ω. This type of rheological behavior is considered as solid-like.

The differences in the parameters of the structure between glycinin and glycinin-P cause differences in the viscoelastic properties of adsorbed layers and the nature of their dependence on protein concentration. Therefore, over the entire frequency range and protein concentrations the observed higher values of E' of adsorption layers are higher in case of glycinin-P. Apparently, more compact and labile structure and a lower charge density of the glycinin-P molecules reduce the energy barrier and form a more perfect adsorption layer, providing a greater number of contacts between the macromolecules. More high surface hydrophobicity of glycinin-P molecules is also favorable for the stabilization of the adsorption layer due to hydrophobic interactions between macromolecules within the layer.

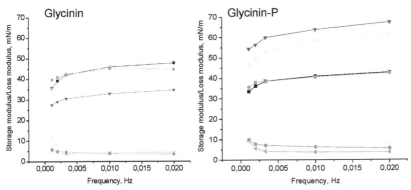

FIGURE 6.2   The real and imaginary components of the dilatation modulus of glycinin/ glycinin-P viscoelastic adsorption layers at different concentrations of Proteins in solution. module of conservation: the blue dots—1.0 mg/ml, green dots—0.1 mg/ml, red dots—0.01 mg/ml, black dots—0.001 mg/ml. Loss modulus: dark green dots—1.0 mg/ml, yellow dots—0.1 mg/ml, pink dots—0.01 mg/ml, blue dots—0.001 mg/ml. Conditions: 0.05 M phosphate buffer, pH 7, 6, 0.5 M NaCl.

6.2   CONCLUSION

Thus, it was found that limited proteolysis with papain leads to a change in glycinin molecular parameters, namely, decrease of molecular mass, radius of gyration, hydrodynamic size and thermodynamic stability. At the same time the surface hydrophobicity of glycinin molecule increases. It is displayed in enhance of the heat capacity increment and in lowing of the second virial coefficient. As a result, a significant growth of its surface activity is observed. Increase of the rate of formation and viscoelasticity of adsorption layers at the air/water interface are caused by decrease in diffusion coefficient, which in turn is caused by decreasing of molecular mass and hydrodynamic size. The interrelation between molecular parameter changes and glycinin surface behavior is established. These changes are favorable for improving of the glycinin functional properties.

KEYWORDS

- **Adsorption layers**
- **Dynamic tensiometry and dilatometry**
- **Glycinin**
- **Hydrodynamic size**
- **Molecular mass**
- **Molecular parameters**
- **Papain**
- **Radius of gyration**
- **Rheology**
- **The limited proteolysis**
- **Thermodynamic stability**

REFERENCES

1. Shutov, A.; et al. Limited proteolysis regulates massive degradation of glycinin, storage 11S globulin from soybean seeds: An *in vitro* model. *J. Plant Physiol.* **2012,** *169,* 1227–1233.
2. Itzhak, R. F.; and Gill, D. M.; "A micro-biuret method for estimating proteins" *Anal. Biochem.* **1964,** *9(4),* 401–410
3. Krivandin, A. V.; Muranov, K. O.; and Ostrovsky, M. A.; Study of complex formation in solutions of α- and β-crystallines at 60 C. *Mole. Biol.* 2004, *38(3),* 532–546.
4. Konarev, P. V.; Volkov, V. V.; Sokolova, A. V.; Koch, M. H. J.; and Svergun, D. I.; PRIMUS: a Windows PC-based system for small-angle scattering data analysis. *J. Appl. Cryst.* **2003,** *36,* 277–1282.
5. Feigin, L. A.; and Svergun, D. I.; Structure Analysis by Small-Angle X-Ray and Neutron Scattering. New York: Plenum Press; **1987.**
6. Ericson, H. P.; Size and shape of Protein Molecules at the Nanometer level, Biological Procedures Online, **2009,** *11,* Number 1.
7. Patino, R.; Carrera, C.; Molina, S.; Rodrıiguez Nino, Ma. R.; and Anon, C.; Adsorption of soy globulin films at the air-water interface. *Ind. Eng. Chem. Res.* **2004,** *43,* 1681–1689.
8. Debrier, J.; and Babak, V. G.; Mini Review. Interfacial properties of amphiphilic natural polymer systems based on derivatives of chitin. *Polym. Int. Polym Int.* **2006,** *55,* 1177–1183.

CHAPTER 7

RUBBERS PROTECTION AGAINST AGING: STRUCTURAL, DIFFUSION, AND KINETIC EFFECTS

V. F. KABLOV and G. E. ZAIKOV

CONTENTS

7.1 INTRODUCTION

Rubber aging is a complicated physical and chemical process running in the nonequilibrium open heterogeneous system. Aging leads to destruction and/or macromolecules structuring, changes in physical structure, running diffusion processes, and other structure and chemical changes in a material.

In this chapter, we shall consider the problem of developing the system technology as a new scientific approach to producing elastomer-olygomer materials. This technology combines general and special theory of systems, thermodynamics, kinetics, physics and chemistry of polymers, and computer methods of data processing.

A large variety of operating environments, a wide range of effects on polymer materials, and the related multitude of technical materials (TMs) require a common conceptual approach to the development of new efficient materials. Such a common conceptual approach is a systematic approach. The system technology combines various complementary methods and approaches to the creation of elastomeric materials. The development strategy based on complementary approaches is seen as the most effective. The main theoretical approaches that underlie the development of materials based on elastomeric systems are given below.

7.2 SYSTEM APPROACH

The task of the system description is a record of all diversity of problems facing both the design and technical material TM as well as finding optimal pair TM product. The system description has to be based on the fundamental objective of materials science, which is to establish relationships between composition, technology, structure, and properties of materials.

The basis of the system approach is the representation of an object as a system, that is, as an integral set of interrelated elements of any nature. The main criterion for establishing the need for this element in the system is its participation in the system operation resulting in obtaining the desired result. The system approach allows for functionally supported dissection of any system into subsystems whose volume and number are determined by the composition of the system and the consideration scope. One might add that the same object can be represented as different systems, while the number of ways of viewing an object using the system approach has no limitations.

Representing an object as a system, we only get the opportunity to approach the structure of the object; a further step is to search for patterns of whole object relations in a system.

One of the systems approach methods is also functional and physical analysis, in which a multicomponent material is considered as a technical system carrying out operations to convert some of the input actions on the output ones. The main attention at that is paid not to the material structure of the object, but to functional transformations of matter and energy flows [2–3].

External flows of matter and energy through functional systems affect a TM and cause its various physical and chemical transformations (PCTs). They lead to a change in the molecular, supramolecular, phase micro and macrostructure of the TM. Since PCT and structural changes occur at different rates, it is necessary to consider the heterochronic behavior of system changes. As a result of all the processes occurring in the TM during operation change their structural and functional characteristics, therefore we can talk about a system change in the TM or TM systemogenesis.

7.3   A THERMODYNAMIC APPROACH

Modern thermodynamic methods can be applied not only to identify the possibility or impossibility of the running processes, but also as methods to find new technical solutions.

A polymer operated at high temperatures, in reactive environments, under intense friction, and other intensive effects can be regarded as a nonequilibrium open system, that is, a system exchanging the medium substance (or energy). Energy and substance exchange is carried out by heat conduction, diffusion of the medium in the material, low molecular additives and products of thermal and chemical destruction of the polymer matrix. Under these conditions, the material as a system is not in equilibrium state, and the further it from equilibrium, the greater the intensity of exposure to the material and, consequently, the more intensive mass and energy exchange with the medium is. The presence of reactive components in a material increases an equilibrium deviation and leads to greater intensification of mass and energy exchange.

For an open system, the total entropy change dS is given by the following equation:

$$dS = d_iS + d_eS,$$ (7.1)

where d_iS is a change (production) in entropy in a system due to irreversible processes and d_eS is the entropy flow due to energy and matter exchange with the medium.

It is important that the entropy of an open system can decrease due to the entropy output in the medium ($d_eS < 0$), and provided that $|d_eS| <> |d_iS|$. Then, despite the fact that $d_iS > 0$ (in accordance with the second law of thermodynamics), $dS < 0$.

The decrease in entropy of an open system means that in such a system self-organization processes start, as well as forming specific spatio-temporal dissipative structures supporting sustainable state of the system.

To ensure the flow of negative external entropy can be implemented as a specially organized by external influences such as the artificially created concentration gradient, temperature, potential, and due to the chemical reactions that lead to high entropy products output of the system and its enrich with low-entropy products (an example is polycondensation system with by-product output).

Aging of elastomeric materials from the thermodynamic viewpoint can be seen as processes of self-organization and disorganization in open non-linear systems [1–2]. Processes occurring in such systems can be operated as an organization of chains of physical and chemical transformations in the systems themselves, highlighting the various functional subsystems as well as by controlled external exposures (thermodynamic forces and flows). In case of exposures, in accordance with the Onsager principle, pairing of thermodynamic forces and flows may take place. This allows for additional opportunities to influence the aging process.

When in a nonequilibrium system several irreversible processes run, they are superimposed on one another and cause a new effect. Irreversible effects may occur due to a gradient of temperature, concentration, the electrical or chemical potential, etc. In thermodynamics all these quantities are called "thermodynamic" or "summarized forces", and denoted by X. These forces cause irreversible phenomena: heat flow, diffusion current, chemical reactions, and other flows, called summarized and denoted by J.

In general, any force can cause any flow that is reflected in the famous Onsager equation:

$$J_i = \sum_{k=1}^{n} L_k X_k \ (i = 1,2,...n) \tag{7.2}$$

Coefficients L_{ik} are called phenomenological coefficients.

Entropy change in time is the sum of products of summarized forces and flows:

$$\frac{dS}{dt} = \sum X_i J_i = \sigma, \tag{7.3}$$

where σ is the entropy production rate per unit volume.

The condition refers to sum $\sum_i X_i J_i$ in general. Certain members of the sum may be negative, which means that separate flows J_i are impossible because $X_i J_i > 0$ (contradiction to the principle of entropy increase). However, when paired with other flows, which correspond to positive values of $X_i J_i > 0$, in an open system a flow inconceivable in a closed system becomes possible. Thus, when the thermal diffusion takes place, the material flow direction is opposite to the concentration decrease direction, since it is linked to the heat flow coming from the hot to the cold wall. A similar process is realized at electrodiffusion. As the system approaches the stationary (but not equilibrium) state, $\frac{d_i S}{dt}$ tends to a minimum value that depends on the given conditions.

If the system deviates from the stationary state, then since the desire to minimize the entropy production according to Prigogine's theorem (extreme principle) about the minimum speed of the entropy production in a stationary state, there come such internal changes that will tend to bring the system to that state. Such an autostabilization phenomenon represents an extension to stationary nonequilibrium systems on Le Chatelier's principle applicable to chemical equilibrium.

A technical task for a developer of physical and chemical systems is to ensure the maximum deviation dA_p from zero in the negative direction, if his purpose is to use the external influences to put the system in a stationary state.

With aging materials near the destruction temperature, the mechanism of aging and, thus, methods of protection can vary significantly [3]. The processes occurring in the intensive thermal destruction, including self-similar and thermal explosion modes, have been observed.

7.4   KINETICS OF PROCESSESS

Kinetics of aging is of quite difficulty. The paper considers a typology of kinetic processes, different types of kinetic curves and their models. It has been shown that in some cases, it is possible to use kinetic effects to prolong performance of elastomeric materials.

An important feature of elastomeric materials is running cooxidation reactions into them. Kinetic features of radical reactions in multicompo-nent polymer systems can lead to accelerated aging of a polymer with a more labile hydrogen atoms and slow oxidation reactions in an inert poly-mer. This factor requires selecting of an anti-aging system.

Under the influence of operational factors on polymer material, it is possible to allocate two modes of exploitation: normal mode and exploita-tion in extreme conditions, for normal one a relatively slow and gradual change of parameters during operation is typical.

Such changes in the parameters are usually described by linear de-pendencies, linear portions of nonlinear dependencies or Arrhenius-type equations. During normal mode of operation, the effective values of the factors (temperature, concentration of corrosive environment, etc.,) are far from the working capacity limits for materials. Under extreme conditions the values of working factors are approaching to the limits of working capacity, or even fall in the "working incapacity." Material in this mode tragically quickly goes out. The change in parameters is usually described by nonlinear dependencies including extreme ones. Material lifetime, however, may be sufficient in some cases—for disposable use, while larger stocks thickness (mass, volume, etc.), as well as in a case of con-structive features of the assembly, the application of protective coatings and other protection methods. Finally, material lifetime can be highly im-proved by the introduction of functionally active ingredients into its com-position protecting the material during operation. Nonlinearity of polymer materials behavior is enhanced as it approaches to the critical operational parameters. Simultaneously, instability of materials increases—structural and dynamic stability. At the same time, nonlinear systems may have the stability zones under certain values of the parameters. Therefore, a system with autocatalysis or autoinhibition can be described by the following equation:

$$\dot{X} = -kX - k_1 X,$$ (7.4)

where $\dot{x}$ is the change in any characteristic, k and k_1 are the–parameters typical for a system.

When $k>0$, such a system has only one real root $X=0$ corresponding to the stationary state, and if $k<0$ it has three real roots, where $X=0$ becomes unstable and the stable ones are two extreme roots $X_{2,3} = \pm\sqrt{\frac{-k}{k_1}}$. It is evident that in passing of these values through 0 to the negative side the state stability is lost, but instead there are two new stable states: there comes a system complication (differentiation) by bifurcation (splitting) of the initial stationary state. Mathematical analysis shows that at sufficiently strong nonlinearity there may be a complication of a structure with the formation of stable states. Complexity of the system structure is even more multivariate in systems with two or more variables.

Systems in a state of instability are particularly sensitive to small external influences that can lead not only to the destruction of the system, but on the contrary, transfer it to a new stable state, which, though not an equilibrium, but has sufficient stability. A typical example is the effect of ultra-low friction, which is appeared in relatively weak irradiation of a rubbing pair "polymer-metal." Another example is the rubber hardening at the initial stages of aging, when a rubber composition contains a polymer forming compounds or structuring additives. Thus, the polymer systems that are in the nonlinear area typical for extreme conditions can be stabilized by internal or external "control" actions. Extreme operating conditions are such as work of heat protective materials in high-temperature gas flows, work of rubber products in the dynamic mode in combination with other actions, and so on.

7.5 A STRUCTURED APPROACH

Since elastomeric materials are microheterogeneous multiphase materials, the great influence on aging have diffusion processes including migration of components. By increasing the operating temperature, volatility of components, including the volatility of antiagers, enhances. Therefore, the urgent problem is reducing the volatility of antiagers while keeping them sufficient lability (ability to "hit" the zones on destructing macromolecules where free radicals appear). The problem is solved effectively enough by the use of composite antiaging systems. We propose the antiaging systems capable of continuous "release" of active labile antiagers of micro- and nanoparticles. Antagers are located in the particles matrix in the supersaturated state. When operating, they continuously migrate into the dispersion medium of an elastomer. Various implementations of this method have been considered.

Structure of elastomeric materials at different levels (supramolecular structures, the structure in filled systems, the presence of several polymer phases, polymer-filler, polymer-fiber, and polymer-polymer interfaces) largely determines the kinetics of aging. In particular, the optimal solution is to provide a continuous phase of aging resistant polymer in the dispersion medium of another one, creating a barrier layers of fillers and protective surface layers and coatings able to withstand static and dynamic loads.

KEYWORDS

- **Aging**
- **Antiagers**
- **Component migration**
- **Cooxidation**
- **Destruction**
- **Elastomers**
- **Kinetic effects**
- **Physical and chemical transformations**

REFERENCES

1. Kablov, V. F.; System Technology of Elastomeric Materials—the Integration Processes in the Development of Materials and Products. Proceedings of XV International Scientific-Practical Conference "Rubber Industry: Raw Materials, Technology". Moscow: Niishp, S. T. C.; et al. May 25–29, **2009**, 4–6.

2. Kablov, V. F.; System Technology of Rubber-Oligomer Compositions. Proceedings of X International Conference on Chemistry and Physical Chemistry of Oligomers "Oligomers-2009". Volgograd; **2009**, 162–191.

3. Kablov, V. F.; Protection of Rubbers against Ageing in the Different Application Conditions with the Use of Structural, Diffusion and Kinetic Effects. Proceedings of XV International Scientific-Practical Conference "Rubber Industry: Raw Materials, Technology". Moscow; Niishp, S. T. C.; et al. May 21–25, **2012**, 5–7.

CHAPTER 8

A CASE STUDY ON STRUCTURING AND ELECTRIC CONDUCTIVITY OF POLYMER COMPOSITES

J. N. ANELI, T. M. NATRIASHVILI, and G. E. ZAIKOV

CONTENTS

8.1 INTRODUCTION

Interest to the processes proceeding at increased temperatures (up to 600 K) in polymer materials is stimulated by a possibility to obtain systems with double conjugated bonds, which exhibits the properties of semiconductors. Intensity of formation of the polyconjugation systems increases with temperature, if pyrolysis proceeds in vacuum or under inert atmosphere [1, 2].

Intramolecular transformations with further change of supermolecular system were studied well on the example of polyacrylonitrile [3]. Thermal transformation of polyacrylonitrile leads to the formation of a polymer, consisted of condensed pyridine cycles with conjugation by C=C bonds, as well as by C=N ones. Concentration of paramagnetic particles increases with temperature of pyrolysis. It is known that the ESR signal is one of the signs of polyconjugation appearance in polymer systems [4]. Deep physical and chemical transformations in polymers proceed at combination of temperature varying with introduction of various donor–acceptor inorganic or organic additives into the reactor.

It is known that one of the ways of obtaining of organic semiconductors is the pyrolysis of low- or high-molecular substances. However, as a rule, this method allows a formation of powder like materials and formation of monolithic ones is connected with additive technological procedures, after which often the material deteriorated.

The main aim of the present work is to obtain pyrolized monolithic materials with wide range of the electric conductivity.

8.2 EXPERIMENTAL

Epoxy resin (ER), novolac phenoloformaldehyde resin (PFR), polymethyl-silsesquioxane (PMS), and fiber glass (FG) were chosen as the initial substances. Pyrolysis of mixtures of the components mentioned, pressed in press-forms, was conducted at various temperature ranged within 500–1,500 K in 10 Pa vacuum. Products obtained in this manner possess good mechanical and electro conducting properties, and are monolythic materials. Pyrolized samples or pyrolyzates were tested by polarization microscopy technique to determine their microstructure. The paramagnetic properties of pyrolizates were investigated using of ESR spectrometer of

Brukker type. The type and mobility of charge carriers investigated were measured by the Hall effect technique.

8.3   RESULTS AND DISCUSSION

Inclusion of fiber glass into compositions was induced by the following idea. It is known [5] that at high temperatures, organosiloxanes react with side hydroxyl groups, disposed on the fiber glass surface. In this reaction, they form covalent bonds with those side groups.

It is known that after high-temperature treatment silsesquioxanes obtain a structure, close to inorganic glass with spheres of regulation due to formation of three-dimension siloxane cubic structures and selective sorption of one of the composite elements is possible on the filler surface in the hardening composite [5].

Figures (8.1) through (8.6) reflect changes of some mechanical, electric, and paramagnetic properties of polymer composites depending on pyrolysis temperature. These dependences are the result of proceeding of deep physicochemical transformations in materials. Combined analysis of the change of microstructure and density of materials (Figure 8.1) with the increase of pyrolysis temperature induces a conclusion that excretion of some volatile fractions of organic part of the material, carbonization of organic residue and caking of glassy fibers cause the increase of pyrolyzate density, based on the composite with polymethylsesquioxane. The limit of pyrolyzate density is reached at temperatures near 1,273 K (Figure 8.1, curve 1), followed by a decrease of the material density due to intensification of thermal degradation processes with pyrolysis temperature increase above 1,273 K.

The material strengthening at elongation extremely depends on the pyrolysis temperature, possessing an intermediate maximum near 1,273 K (Figure 8.1, curve 2). Burning out of organic part of the composite leads to weakening of adhesion forces in the interphase and, consequently, to decrease of the material strengthening with pyrolysis temperature increase up to a definite value. At further increase of pyrolysis temperature on the curve of this dependence, the small maximum appears due formation of covalent chemical bonds between glass and organic conjugated double bounds skeleton. At more high temperatures of pyrolysis, the degradation of these bonds has place.

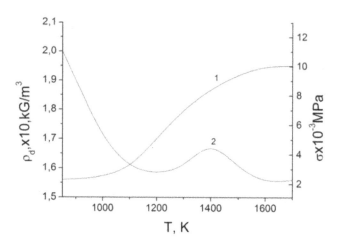

FIGURE 8.1   Dependences of ρ_{den} (1) and strengthening at elongation σ and (2) on pyrolysis 5emperature for the composite ED-20 + PFR + KO-812 + FG.

The conductivity (γ) and charge carrier mobility of the pyrolyzates grows monotonously initially with increasing of the pyrolysis temperature and then saturate. This dependence points out a constant accumulation of polyconjugation systems due to complex thermochemical reactions. Chemical bonds that link organic and inorganic parts of the composite reliably increase stability of polyconjugated structures, responsible for electrically conducting properties of materials. The electrically conducting system of the materials can be considered as a heterogeneous composite material, consisted of highly conducting spheres of polyconjugation and barrier interlayers between them.

The most apparently true model of electric conductivity in materials with the system of double conjugated bonds seems to be the change transfer in the ranges of polyconjugation possessing metal conductivity and jump conductivity between polyconjugation spheres [6].

An important information on the nature of conductivity of pyrolized polymer materials is given by investigation of the γ dependence on temperature. Comparison of the experimental data on dependence of γ-T with known for organic semiconductors [2]:

$$\gamma = \gamma_0 \exp\left(-dE/kT\right) \tag{8.1}$$

and one proposed by N. Mott shows that the dependence obtained by us experimentally satisfies to Mott low [7]:

$$\gamma = \gamma_0 \exp\left[-\left(\frac{T_0}{T}\right)^{1/4}\right],$$ (8.2)

where T_0 and γ_0 are constants depending on some quantum-mechanical values (Figure 8.2).

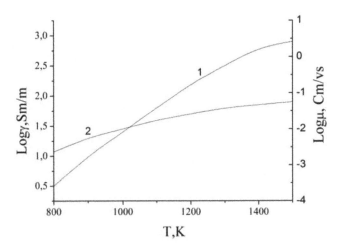

FIGURE 8.2   Dependences of electric conductivity γ (1), mobility of charge carriers μ and (2) on pyrolysis temperature for the composite ER + PFR + PMS + FG.

The growth of carrier mobility μ is well described with analogical expression (Figure 8.3):

$$\mu = \mu_0 \exp\left[-\left(\frac{T_0}{T}\right)^{1/4}\right]$$

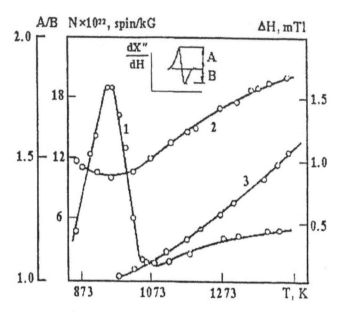

FIGURE 8.3   Temperature dependence of paramagnetic centers N (1) ESR line width, (2) ESR line asymmetry parameter A/B, and (3) on the pyrolysis temperature for the ED-20 + PFS + KO-812 + FG composites.

The dependence of paramagnetic centers concentration in pyrolized polymer composites on pyrolysis temperature has an extreme character. Curve of the present dependence possesses maximum, which is corresponded to the 900–1,000 K range. Change of the ESR absorption line intensity is accompanied by a definite change of its width. In this case, the form and width of the ESR line changes (at constancy of the g-factor)—lines are broadened, and asymmetry of singlet occurs. Maximum on the concentration dependence for paramagnetic centers on pyrolysis temperature is correspondent to the temperature range, in which volatile products of pyrolysis are released and polyconjugation systems occur. Decrease of concentration of the centers above 973K proceeds due to coupling of a definite amount of unpaired electrons. According to this coupling new chemical structures occur (e.g., polyconjugation responsible for electric conductivity increase).

At more high temperatures of pyrolysis deepening of thermochemical reactions in composites leads to formation of the paramagnetic centers localized on the oxygen atom. On the contrary, it is probable of the increase of free charges-current carriers contribution into ESR signal, the line of which is characterized by asymmetry (the so-called Dayson form [8]).

8.4 CONCLUSIONS

High-temperature treatment (pyrolysis) of polymer composites in the inert atmosphere or in the hydrogen medium stimulates processes of formation of the polyconjugation systems). Charge transfer between polyconjugation systems is ruled by the jump conductivity mechanism with variable jump length. In this case, its temperature dependence is described by the Mott formulas. Presence of a glassy fiber and polymethylsilsesquioxane in composites promote formation of covalent bonds between organic and inorganic parts of the composite at pyrolysis. This leads to improving of mechanical properties of materials together with the electric ones.

KEYWORDS

- **Charge transport**
- **Electric conductivity**
- **Paramagnetic centers**
- **Phenolformaldehide and epoxy resins**
- **Pyrolysis**
- **Silicon organic compound**

REFERENCES

1. Brütting, W.; Physics of Organic Semiconductors. Wiley-VCH; **2005**.
2. Aviles, M. A.; Gines, J. M.; Del Rio, J. C.; Pascual, J.; Perez, J. L.; and Sanchez-Soto, P. J.; *J. Thermal Anal. Calorimet.* **2002**, *67*, 177–188.
3. Fialkov, A. S.; Carbon-Graphite Materials. Moscow: Energia; **1979**, *158*.
4. Milinchuk, V. K.; Klinshpont, E. R.; Pshezhetskii, S. Y.; Macroradicales. Moscow: Khimia; **1980**, *264*.
5. Aneli, J. N.; Khananashvili, L. M.; and Zaikov, G. E.; "Structuring and Conductivity of Polymer Composites". New York: Nova Science Publication; **1998**, 326 P.
6. Kajiwara T.; Inokuchi H.; Minomura S.; *Jap. Plast. Age.* **1974**, *12(1)*, 17–24.

7. Mott N. F.; and Davis E.; Electron Processes in Noncrystalline Materials. 2nd Ed. Oxford: Clarendon Press; **1979**.

8. Dyson F. J.; *Phys. Rev.* **1955**, *98*, 349–358.

CHAPTER 9

A STUDY ON MAGNETIC PROPERTIES OF RB$_3$C$_{60}$ SINGLE-CRYSTAL FULLERENE

J. CHIGVINADZE, V. BUNTAR, G. ZAIKOV, S. ASHIMOV, T. MACHAIDZE, and G. DONADZE

CONTENTS

9.1   INTRODUCTION

The appearance of superconductivity in alkali fullerides has led to extensive efforts in attempting to understand their electronic, magnetic, structural, and dynamic properties and to elucidate the origin of their high T_c. In particular, the question of whether or not such a large value of T_c can be caused by coupling to phonons alone is yet to be answered, and the answer strongly depends on the normal-state properties. Despite the apparent simplicity of the structure of A_3C_{60} fullerides (where A is an alkali metal), some important issues are yet to be fully resolved.

It is well known that A_3C_{60} fullerides, as well as pure C_{60}, are weak diamagnetics with the appearance of strong diamagnetism at the transition to the superconducting state. At high temperature, solid A_3C_{60} forms a face-centered cubic (fcc) phase. In this phase, C_{60} molecules freely rotate with a reorientation time scale of the order of 10^{-11} sec. With lowering temperature, the transition to sc structure occurs, which comes from the fast reorientation of the C_{60} molecules, and an anisotropic uniaxial rotation of fullerene molecules takes place. This transition has been observed for K_3C_{60} by NMR experiments [1, 2], but the exact transition temperature was hard to evaluate since, in these experiments, the peak related to this transition overlapped with a peak related to appearance of T` site and is very wide. No special effect on the magnetic properties at the temperature of the solid-state transition from sc to fcc structure has been observed.

A_3C_{60} fullerides are extremely air sensitive and usually are sealed in a glass or quartz capsules. Therefore, most of the magnetic measurements on these materials have been done with SQUID magnetometers. Therefore, an alternative method of investigation of the magnetic properties may give new information about the magnetic structure of A_3C_{60} superconductors in the normal state.

9.2   EXPERIMENTAL METHOD AND SAMPLES

In these experiments, we apply the mechanical torque method to study the magnetic properties of alkali-doped fullerenes. This method has been extensively used to investigate critical parameters of superconductors such as T_c, critical field H_{c2} and energy dissipation in the mixed state [3, 4]. A cylindrical (in the ideal case) sample is suspended on a thin elastic thread and torque oscillations of small amplitude of $1°$ or $2°$ are generated with a short-time impulse. After that, the sample performs free axial-torsional oscillations in an external magnetic field, $\vec{H}$, which is perpendicular to the axis of the sample. The temperature dependence of the frequency, ω (or of the period, t), and of the dissipation of the oscillations, δ, is measured at different magnitudes of the magnetic field.

If there are no fixed (pinned) magnetic moments in the sample, neither the dissipation nor the frequency of the oscillations depends on the external magnetic field. For example, when either (i) the external magnetic field does not penetrate the substance, which is the case for a superconductor in an external field smaller than the lower critical field H_{c1}, or (ii) the inner magnetic moments are either zero or disoriented and not fixed.

The appearance of pinned magnetic dipoles produces a nonzero magnetic moment $\vec{M}$ in the sample. The interaction between $\vec{M}$ and $\vec{H}$ makes a torque $\tau = MH \sin\alpha$, where α is the angle between $\vec{M}$ and $\vec{H}$. This additional moment τ affects the oscillating system and makes the dissipation and the frequency of oscillations dependent on the external magnetic field. The sensitivity of this method is very high, 10^{-17} W [4].

In our experiments, we used a crystal of Rb3C60 that was made from a single crystal of C_{60} by doping it with Rb using the method of vapor-phase doping (Sample preparations have been done in the Institute for Material Physics, Vienna University, Vienna, Austria.). Details of the sample preparation and its characterization can be found in Refs. [5, 6] (sample R28 therein). The physical dimensions of the crystal are 3.3 × 2.7 × 1.3 mm³. It consists of a few single-crystalline grains of the radius about 350 μm [5]. The sample is sealed in a quartz capsule to prevent exposure of the material to air.

9.3   RESULTS AND DISCUSSIONS

In Figure 9.1, we present temperature dependences of both period and dissipation of the oscillations of the sample monitored in the external magnetic field of 100 mT with increasing temperature from the temperature of liquid He. At temperature T_c = 28 K, there is a step-like transition to the normal state on both t and δ. This transition temperature is slightly lower than that obtained in Ref. [6] for this crystal. This difference is because the thermometer with the electrical cables cannot be fixed on the oscillating sample directly and is placed some distance from it.

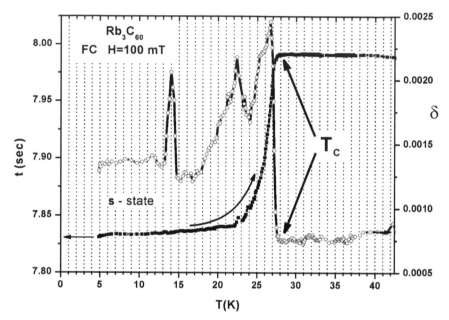

FIGURE 9.1 Temperature dependence of the period (solid squares) and of the dissipation of the oscillations (open circles) of Rb_3C_{60} in the superconducting state and at temperatures close to the transition, T_c, in the external magnetic field $\mu_0 H = 100$ mT.

It can be clearly noted that the dissipation in the superconducting state is larger than that in the normal state. This is because, in the superconducting state, the material is penetrated by magnetic field in the shape of vortices. These vortices are partly pinned on structural defects, and the pinning force depends on several factors such as size and number of the defects, and thermal fluctuations. Vortices in the material tend to align along the external magnetic field, which increases the dissipation of the oscillations. When the material undergoes the transition to the normal state, vortices do not exist in the sample anymore and, therefore, do not prevent the oscillation of the sample.

The larger the amplitude of the oscillations, the larger the force applied on a pinned vortex by the magnetic field. By measuring the period t and the dissipation δ at different amplitudes of oscillations, one can find the critical amplitude φ_c at which vortices unpin from structural defects. Using the value of φ_c one can find the value of the critical moment τ_c and the strength of the bulk pinning force F_p as [7]:

$$F_p = \frac{3}{4} \frac{\tau_c}{R^3 L};$$

(1)

where R and L are the radius and the length of the cylindrical sample, respectively. The results of the estimation of the bulk pinning force at temperature $T = 4.2$ K are presented in Figure 9.2.

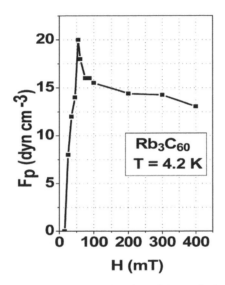

FIGURE 9.2 The bulk pinning force as a function of the applied external magnetic field at $T = 4.2$ K.

With the increasing external magnetic field from zero, vortices penetrate the sample from its surface to the center and the bulk pinning force grows steeply. At the external field $H_{ext} = H^*$ (in our experiments $H^* \sim 50$ mT), the sample is completely penetrated by vortices. Further increasing of the external magnetic field leads to increasing numbers of vortices, (i.e., the volume density of vortices in the sample), and therefore to the decreasing of the bulk pinning force.

The temperature dependence of the dissipation of the oscillations in the normal state is presented. At around $T_1 \sim 270$ K, a peak can be observed. This peak may be associated with the first-order phase transition from sc to fcc structure. This transition is related to the freezing of libration modes and an orientation of C_{60} molecules. If each C_{60} molecule is considered as a diamagnetic dipole, then the orientation of molecules leads to the orientation of magnetic moments in the system and, therefore, strong response on $\delta(T)$ and $\omega(T)$. As expected, the transition temperature in Rb_3C_{60} is higher than that in the pristine C_{60} at $T = 263$ K due to the presence of Rb atoms in the inter fullerene space.

There is a second peak at $T_2 \sim 200–250$ K. The amplitude of the peak of the dissipation (as shown in Figure 9.3) is several orders of magnitude, which is above of the sensitivity of our experimental device. Moreover, this peak is much larger than the one at the sc–fcc phase transition. In our experiments, the sample stops oscillating, and we cannot generate the oscillations with a short-time impulse any more. We are not aware of any earlier experiments showing such strong magnetic response of fulleride materials.

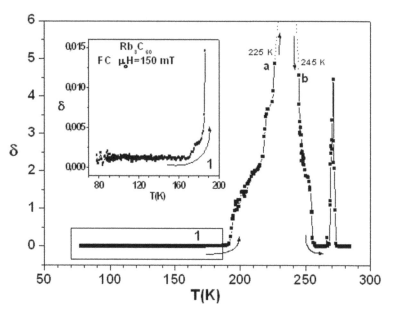

FIGURE 9.3 Temperature dependence of the dissipation δ of the oscillations at the external magnetic field $\mu_0 H = 150$ mT.

In earlier NMR experiments, some anomalies of the temperature dependence of spin-lattice relaxation times have also been observed in this range of temperatures in K_3C_{60} [8, 9] and Na_2CsC_{60} [10]. However, this anomaly was very weak and, as the authors reported, its amplitude was in the order of experimental error. In our experiments, the magnitude of the effect is huge. Since the effect appears to be so strong in our torque experiments, which are sensitive to the presence of fixed/oriented magnetic moments, it may mean that at this temperature there is a reorganization of the magnetic structure in the material.

In is unclear now as to what kind of magnetic structure is formed in A_3C_{60}. These magnetic moments cannot be the moments of C_{60} molecules since most of the molecules are already oriented at $T_1 = 280$ K and, moreover, the magnetic effect is much stronger than that at the phase transition. One may speculate that some persistent currents may appear, or we may assume that the magnetic moments involved in the crossover are due to distortions of the C_{60} molecules or the A_3C_{60} lattice. In general, magnetism is mainly related to properties based on electron spins. The solid lattice, though important for some of effects, usually does not provide an important contribution to magnetism. However, there is a specific type of materials for which the Jahn–Teller effect plays a very important role determining both structural and magnetic properties.

KEYWORDS

- **Magnetic state**
- **Normal state**
- **Rb_3C_{60} fullerene**
- **Superconductivity**

REFERENCES

1. Yoshinari, Y.; Alloul, H.; Kriza, G.; and Holczer, K.; *Phys. Rev. Lett.* **1993**, *71*, 2413.
2. Yoshinari, Y.; Alloul, H.; Brouet, V.; Kriza, G.; Holczer, K.; and Forro, L.; *Phys. Rev. B*, **1996**, *54*, 6155.
3. Chigvinadze, J.; *JETP.* **1972**, *63*, 2144.
4. Chigvinadze, J.; *JETP.* **1973**, *65*, 1923.
5. Buntar, V.; Haluska, M.; Kuzmany, H.; Sauerzopf, F. M.; and Weber, H. W.; *Supercond. Sci. Technol.* **2003**, *16*, 907.
6. Buntar, V.; Zauerzopf, F. M.; Weber, H. W.; Halushka, M.; and Kuzmany, H.; *Phys. Rev. B*, **2005**, *72*, 024521.

7. Fuhrmans, M.; and Heiden, C.; *Cryogenics.* **1976,** *8,* 451
8. Matus, P.; Alloul, H.; Kriza, G.; Brouet, V.; Singer, P. M.; Garaj, S.; and Forro, L.; *Phys. Rev. B,* **2006,** *74,* 214509.

CHAPTER 10

MODIFICATION OF ETHYLENE COPOLYMERS BY AMINOALKOXY-, GLYCIDOXYALKOXYSILANES

N. E. TEMNIKOVA, S. N. RUSANOVA, S. YU. SOFINA,
O. V. STOYANOV, R. M. GARIPOV, A. E. CHALYKH,
V. K. GERASIMOV, and G. E. ZAIKOV

CONTENTS

10.1   INTRODUCTION

Polyethylene and copolymers of ethylene with vinylacetate (EVA) and from the beginning of the twenty-first century copolymers of ethylene with vinyl acetate and maleic anhydride (SEVAMA) are commonly used as the main ingredients of hot melt adhesives, which determine the cohesive strength, heat resistance, processing conditions, and the use of the composition. It is known that aminoalkoxysilanes and organosilanes containing glycidoxy groups are widely used in fiberglass and paint industries, improving the adhesion of different polymers and coatings (acrylic, alkyd, polyester, and polyurethane) to inorganic substrates (glass, aluminum, steel, etc.). Therefore, increase of the adhesive strength of the compositions based on polyolefins and these organosilanes to various substrates (steel, aluminum) is to be expected.

Earlier [1–5], we studied the changes in the chemical structure of ethylene copolymers during organosilane modification and the effect of organosilicon compounds on the structural, mechanical, operational, and technological characteristics of the modified copolymers. The purpose of this research work was to study the adhesion characteristics of ethylene copolymers modified by aminoalkoxy- and glycidoxyalkoxysilanes.

10.2   SUBJECTS AND METHODS

Copolymers of ethylene with vinyl acetate (EVA) Evatane 20-20 and Evatane 28-05; copolymers of ethylene with vinyl acetate and maleic anhydride (SEVAMA) grades Orevac Orevac 9,305 and 9,707 were used as the objects of the study. The main characteristics of the polymers are listed in Table (10.1). Modifier—(3-glycidoxypropil) trimethoxysilane. The reaction mixing of copolymers with modifiers was conducted on laboratory roll mills at a rotational speed of rolls 12.5 m/min and friction ratio of 1:1.2 for 10 min in the range of 100–120°C. Modifier content was varied in the range 0–10 wt. %. Adhesive joint strength of the samples was evaluated by flaking method the day after forming an adhesive compound in accordance with GOST 411-77 (to metal substrates) and GOST 2896. 1-91 (to PET). Metal substrates: steel 3 (St3) and aluminum AMG-1.

TABLE 10.1 Characteristics of ethylene copolymers

Brand of EVA	EVATANE 20-20	EVATANE 28-05	Orevac 9305	Orevac 9307
Unit designation	EVA20	EVA27	SEVAMA26	SEVAMA13
VA content (%)	19–21	27–29	26–30	12–14
MA content (%)	–	–	1.5	1.5
Melt flow rate, r/10 min, $t =$ 190°C, load 5 kg	59.29	18.16	92.77	26.76
Melt flow rate, r/10 min, $t =$ 125°C, load 2.16 kg	2.67	0.76	14.72	1.02
Density (kg/m³)	938	950	951	939
Breaking stress at tension (Mpa)	14	23.88	7	19.5
Elongation at break (%)	740	800	760	760
Elastic modulus (Mpa)	31	17	7	61

10.3 RESULTS AND DISCUSSION

The interaction of functional groups of adhesive and substrate was considered in many studies, but the optimal content of active groups in the adhesive is often selected empirically because in most cases there is no proportionality between the adhesive strength and the content of the functional groups in adhesive. According to Ref. [6], this dependence is often extreme, or with the increase of the content of the functional groups adhesive strength reaches a certain limit and no longer increases.

It was found that the introduction of 2 per cent by mass of AGM-9 increases the adhesion strength to steel (St3) for EVA27 in four times and for SEVAMA26 in 1.75 times. The further introduction of the modifying agent does not lead to an increase in the strength of the adhesive compound (Figure 10.1).

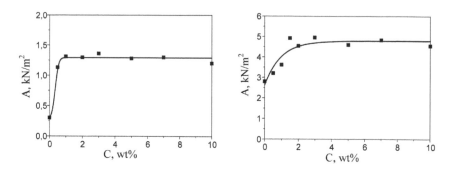

FIGURE 10.1 Adhesive strength polymer—steel (St3): (a)—EVA27-AGM-9; (b)—SEVAMA26-AGM-9. Condition of formation 160°C, 10 min.

It was also found that the introduction of up to 0.5 wt. % DAS and AGM-9 increases the adhesion strength to aluminum (AMG-1) for the copolymers EVA20-AGM-9 in 1.75 times and for SEVAMA26-DAS in 14 times. The further increase of the modifying agent concentration reduces the adhesion strength (Figure 10.2) relatively to the maximum value.

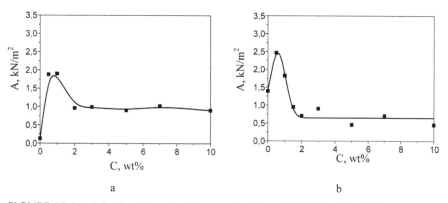

a   b

FIGURE 10.2 Adhesive strength polymer—aluminum (AMG-1): (a)—EVA20-AGM-9; (b)—SEVAMA26-DAS. Condition of formation 160°C, 10 min.

For terpolymers destruction of the system adhesive—metal with the modifier content up to 2 per cent has the mixed nature; a bundle of the adhesive coating takes place, because adhesive joint strength of the system polymer—metal is higher than the strength of the adhesive. In this case, the adhesive strength of the modified and nonmodified terpolymers is significantly higher than that of the double copolymers owing to the presence of maleic anhydride in SEVAMA. At a modifier concentration of more than 2 per cent the gap is purely adhesive.

Modification of ethylene with vinyl acetate copolymers by glycidoxyalkoxysilane has no effect on the strength of the adhesive contact ethylene copolymer—PET. However, the introduction of 1.5 wt. % of this additive in SEVAMA leads to a significant (in 3.4 times) increase in adhesion strength of the system, and the destruction of the substrate is observed (Figure 10.3).

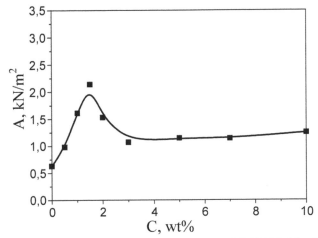

FIGURE 10.3 Adhesive strength polymer—PET: SEVAMA26-GS. Condition of formation 160°C, 10 min.

Adhesive strength of the system polymer—substrate depends on the nature of the contacting materials (their surface energy) and the conditions of the adhesive contact formation.

Table (10.2) shows the experimental data on the surface energy and its polar and dispersion components for all grades of the tested copolymers. The surface energy of the initial copolymers is determined mainly by the dispersion component.

TABLE 10.2 Energy characteristics of the copolymers

Grade	γ (mJ/m²)	γ^D (mJ/m²)	γ^P (mJ/m²)
EVA20	36	25.1	10.9
EVA27	38.6	25	13.6
SEVAMA13	40.4	27.9	12.5
SEVAMA26	37.4	28.4	9

These data were applied on the generalized temperature-concentration diagram of the surface energy of EVA (Figure 10.4), taken from Ref. [7], which includes both the operating conditions and testing of adhesive joints (curve I), and also the conditions of the adhesive joints formation (curve II). Traditionally [8], dependencies of the surface energy from the composition of the random copolymers describe by the simple additive function—the dotted lines. As can be seen in Figure (10.4), the surface energies of the copolymers are close to additive values.

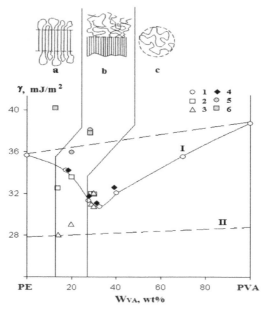

FIGURE 10.4 Dependence of the surface energy on the composition of EVA copolymers. The samples were formed in contact with: air—1 [9], 2, 4 [8], PTFE—3[10]. 5—EVA, (PET), 6—SEVAMA (PET).

In Ref. [10], it was shown that the total surface energy of the copolymers is a function of the substrate surface nature, in contact with which the surfaces of the copolymers samples were formed. Typically, the surface energy of the samples obtained on high-energy surfaces (steel, glass, aluminum, and polyethylene PET) is significantly higher than the surface energy of the samples prepared in contact with low-energy surfaces (PTFE, its copolymers, and PE), and in the air. Since the formation of the samples was carried out on PET, this explains the higher values of the surface energy for the studied copolymers, relative to the copolymers formed in contact with low-energy surfaces.

Thus, on the basis of these results in Ref. [7] concluded that, when exposed to high-energy substrates in the surface layers of the copolymers conformational changes take place, which result in changes of vinylacetate units concentration, on the one hand, and in changes in the packing density of macromolecules segments on the other hand. For high-energy substrates, the number of vinylacetate units in the surface layers of the copolymers is close to their number, characteristic of the surface of PVA. Formation of the adhesive compound modified copolymer—substrate was carried out in contact with the high-energy substrates. Unreacted groups of VA along with grafted siloxane concentrate on the composition surface. Both of these factors, complementing each other, lead to increase of adhesion between composition and steel (aluminum and PET).

10.4   CONCLUSION

Modification of ethylene *copolymers* by aminoalkoxy- and glycidoxyalkoxysilanes increases the strength of the adhesive contact both to metals (steel and aluminum) and polar polymers (PET). Therefore, these materials having enhanced deformation, strength and adhesive properties may be used in the production of multilayer polymer films, including metalized polymer films.

ACKNOWLEDGMENT

The work was supported by the Russian Ministry of Education as a part of a comprehensive project under the contract no 02.G25.31.0037, according to the decision of the Government of the Russian Federation no 218 of April 9, 2010.

KEYWORDS

- **Adhesion**
- **Ethylene copolymers**

REFERENCES

1. Rusanova, S. N.; Diss. ... kand. tekhn. nauk: 02.00.16/Rusanova Svetlana Nikolaevna. Kazan: Kazanskii Gos. Tekhnol. Un-t; **2000**. 119 s.
2. Temnikova, N. E.; Vliyanie aminosoderzshashego modificatora na svoistva sopolimerov etilena. Ed. Temnikova, N. E.; Rusanova, S. N.; Tafeeva, Yu. S.; Sofina, S. Yu.; Stoyanov Klei, O. V.; Germetiki. Tekhnologii; **2012**, *4*, S, 32–48.
3. Rusanova, S. N.; Modifikatsiya sopolimerov etilena aminotrialkoksisilanom. Ed. Rusanova, S. N. ; Temnikova, N. E.; Mukhamedzyanova, E. R.; Stoyanov, O. V.; Vestnik Kazanskogo Tekhnologicheskogo Universiteta; **2010**, *9*, S, 353–355.
4. Rusanova, S. N.; IK-spektroskopicheskoe issledovanie vzaimodeistviya glitsidoksisilana i sopolimerov etilena. Rusanova, S. N.; Temnikova, N. E.; Stoyanov, O. V.; Gerasimov V. K.; Chalykh A. E.; Vestnik Kazanskogo Tekhnologicheskogo Universiteta; **2012**, *22*, S, 95–97 p.
5. Temnikova, N. E.; Issledovanie modificatsii sopolimerov etilena aminosilanami metodom IK-spektroskopii NPVO. Temnikova, N. E.; Rusanova, S. N.; Tafeeva, Yu. S.; and Stoyanov, O. V; Vestnik Kazanskogo Tekhnologicheskogo Universiteta; **2011**, *19*, S, 112–125.
6. Currat, C.; Silane croslinker insulation for medium voltage power cables. Currat, C.; *Wire J. Int.* 1984, *17(6)*, 60–65.
7. Gladkikh, Yu. Yu.; Deformatsionno-prochnostnye i adgezionnye svoistva sopolimerov etilena i vinilatsetata: Diss. ... kand. khim. Nauk. Gladkikh Yuliya Yurievna. Moskva: IFKhE RAN; **2012**, 156 s.
8. Vu S.; Mezshfaznaya energiya, struktura poverkhnostei i adgeziya mezshdy polimerami. S. Vu v sb.: Polimernye smesi pod red. Pola D. i Niumena S. v 3 t. T.1. M.: Mir; 1981, S, 282–332.
9. Nose, T.; Theory of polymer liquids and glasses. Nose, T.; and Hole, A.; *Polym. J.* **1972**, *l.3(1)*, 1–11.
10. Balashova, E. V.; Vliyanie predistorii na poverkhnostnye svoistva polimerov v razlichnykh fazovykh sostoyaniyakh: Diss. ... kand. khim. nauk/Balashova Elena Vladimirovna. Moskva: IFKhE RAN; **2003**, 155 s.

CHAPTER 11

A RESEARCH NOTE ON NEW FILM-FORMING COMPOSITIONS ON THE BASIS OF EPOXY RESIN AND INDUSTRIALLY PRODUCED ISOCYANATE

N. R. PROKOPCHUK, E. T. KRUTS'KO, and F. V. MOREV

CONTENTS

11.1 INTRODUCTION

Epoxy and polyurethane materials have superior performance properties, so they are used for the production of high-quality coatings [1]. Each of the above-mentioned types of polymers has its advantages and disadvantages: epoxies have low shrinkage during hardening, high chemical resistance, hardness, adhesion to polar surfaces, and high dielectric performance, but they are inferior to polyurethane materials in resistance to aromatized fuel, abrasion resistance, and adhesion to aluminum and nonferrous metals. On the other hand, polyurethanes have limited resistance to alkalis and acids and they are inferior to epoxy resin in strength and hardness.

There are different ways of mutual modification of both types of polymers, which are nowadays presented in numerous publications, mostly in the patents, the number of which is progressively increasing. Analyzing the available references on the subject, we can identify the following ways of obtaining epoxy-urethane coatings: coatings obtained with the use of epoxy-urethane oligomers; coatings produced by hydroxyl-containing epoxy oligomers, isocyanates and their adducts, coatings produced by oligoepoxides, polyisocyanates, and other reactivity compounds, which are coatings produced without the use of isocyanates. In order to improve the chemical resistance as well as the elasticity of epoxy resin composition, blocked isocyanates were used on the basis of the epoxy resin ED-20 as a modifier [2]. As shown by the authors, the reaction of an epoxy resin with an isocyanate at a high temperature results in the formation of heat-resistant poly-oxazolidones. This gives hardness to the coating, improves its mechanical and electrical properties, and urethane linkage of the polymer coating, formed alongside with poly-oxazolidones, making it elastic.

It is known that by varying the chemical structure of hydroxyl component and polyisocyanate, we can alter significantly the hardening process conditions (temperature, rate of drying) as well as performance characteristics of the coatings. Owing to the diversity of the chemical structure of the raw materials, the properties of anticorrosion coatings of paints and lacquers can vary widely. In order to study the properties of the resulting composite film-forming systems on the basis of epoxides and polyisocyanates, it is necessary to study the effect of the quantitative and qualitative structures of the film-forming epoxy-isocyanate compositions on the physical and mechanical properties of coatings.

11.2 MAIN PART

This chapter presents the results of complex study of performance characteristics of coatings on the basis of epoxy resin modified by aliphatic polyisocyanate promising to protect steel constructions from corrosion in a humid atmosphere. It is known that an intense destruction of the polymer coating and corrosion of protected metal occur during the use under various atmospheric conditions.

In this connection, the evaluation of the coatings resistance under atmospheric conditions and the prediction of their service life is an important task.

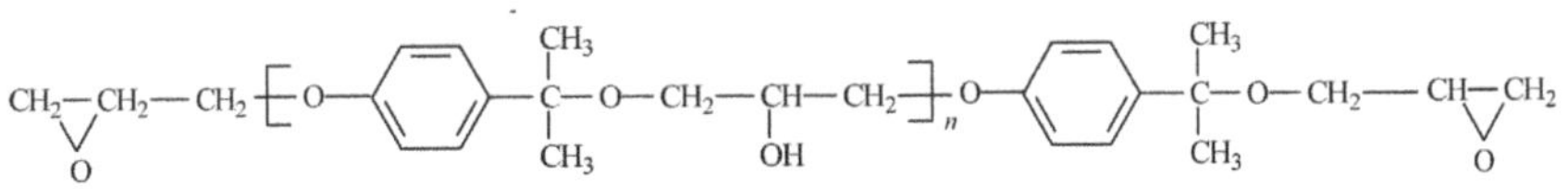

Scheme 1

Epoxy brand ED-20 was used as the object of study (Scheme 11.1), the characteristics of which are presented in Table (11.1).

TABLE 11.1   Properties of epoxy resin ED-20

Index	Value
Average molecular weight	390–430
Epoxy group content, per cent	19.9–22.0
Hydroxyl content, per cent, not more	1.7
Density at 20 ° C, kg/m^3	1,166
Viscosity at 20°C, Pa • s	13–28
Volatile substances content, per cent, not more	1.0

It is known that aliphatic isocyanates are less prone to yellowing under UV light; this is the reason why they are preferred in varnish compositions. In this context, commercially produced aliphatic polyisocyanate 2 K 100 was chosen as a modifier. Polyethylene polyamine was used as a hardener (TU 6-02-594-85).

Film-forming composites were obtained by adding a modifier in the amount of 1–5 per cent by weight of dry matter to epoxy ED-20 resin and hardener (PEPA), followed by stirring the mixture to obtain a homogeneous mass. Films on the metal (copper and steel) and glass substrates were cast from the obtained lacquer solutions. The coatings were made with the help of hydraulic spraying. Plates with formed coatings were stored for 7 days at a temperature of 20–26°C before testing.

Impact resistance of coating samples was evaluated using the device U1 according to the standard ISO 6272 and State Standard 4765-73. The method of determining the strength of films being hit is based on the instantaneous deformation of the metal plate with a varnished coating during the load-free fall on the specimen.

Hardness of a varnished coating was determined with the use of a pendulum device (ISO 1522). The essence of the method is to determine the decay time (the number of oscillations) of the pendulum in contact with its varnished coating.

Adhesive strength of the formed coating was determined in accordance with the standard method as well as ISO 2409 and State Standard 15140-78 using the cross-cut method. The essence of the method is to apply cross-cut of the finished coating, followed by a visual assessment of the state of the lattice coating.

Flexural strength of the coatings was determined using the device SHG1 (ISO 1519, State Standard 6806-73). To perform the determination, a coated sample is bending slowly around the test cylinder, starting with a larger diameter at an angle of 180°. At one of the diameters of the cylinder, a coating either cracks or breaks. In this case, we assume that the coating has an elasticity of the previous diameter of the test cylinder device in which it is not destroyed.

In epoxy-isocyanate catalyst systems without a catalyst, the main reaction at temperatures below 60°C is the reaction of urethane formation through the interaction of isocyanate and secondary hydroxyl groups of epoxy-oligomers (Scheme 11.2).

In cases when the coating formation occurs *in vivo* (without heat), the formation of cross-linked polymer is complicated by the fact that the reactivity system can go into the glassy state, so that the hardening reaction practically stops. Thus, low-temperature hardening of epoxy composites does not enable us to produce coatings with high performance. Another significant drawback of such hardening is its duration. Therefore, the formation of coatings was performed not only under natural conditions, but also at elevated temperatures. To do this, we have chosen two heat settings: 7 days without heat (temperature I) and 2 h at 100°C (temperature II).

$$\text{ww } O-CH_2-CH-CH_2\text{ ww} + O=C=N\text{www} \longrightarrow$$
$$\overset{|}{OH}$$

$$\longrightarrow \quad \text{www } O-CH_2-\overset{|}{CH}-CH_2 \text{ www}$$
$$\overset{|}{O}$$
$$\overset{|}{C}=O$$
$$\overset{|}{NH}\text{ www}$$

To determine the optimal ratio of epoxy oligomer—modifier to perform complex physical and mechanical tests of coatings with different contents of the modifier, a number of physical and mechanical tests of coatings with different contents of the modifier were performed. The results are given in Table (11.2). As seen from Table (11.2), the bending strength reaches a maximum when the content of the modifier is 2–3 per cent. Impact strength also increases with the increase in the modifier content in the composition of the selected concentration range modifier. In this case, the hardness of the coating remained at an acceptable level for the operating conditions used in lacquer coatings.

Physical and mechanical properties of the coatings formed at different temperatures of hardening differ significantly. Increasing hardening temperature affects the concentration dependences of physical and mechanical characteristics of the coatings over the whole range of the investigated ratios of components. Thus, the coating containing the same amount of modifier and cured at an elevated temperature have higher relative hardness, impact strength, and elasticity of less than coatings formed without heat. Thus, the coatings containing the same amount of modifier and hardened at an elevated temperature possess higher relative hardness, impact strength, and less elasticity than those formed without heat. Such changes in the behavior of the physical and mechanical properties of the composites can be explained by a significant change in the level of molecular mobility and packing density in their transition to more "hard" conditions of hardening. However, increasing the hardening temperature does not change the general character of the modifying effect, but only determines its value.

According to the adsorption theory, the adhesive strength of the coatings is caused by the formation of physical and chemical bonds between the macromolecules and the active sites of the solid surface. The observed increase in the adhesive strength of the film-forming compositions developed by us apparently is connected with the flexibility of the emerging polymer grid, which contributes to a more favorable arrangement of the polymer chains with respect to adhesion-active sites of the substrate. We should also note the contribution to the improvement of the adhesive strength of urethane groups capable of forming bonds of the coordination type with a metal surface.

Internal strains occurring in the process of coatings formation and resulting in appearance of local connections between structural elements and adsorption interaction of a film-forming substrate with the service affect the adhesion strength of the film to the metal. With the introduction of the modifier, the flow of relaxation processes is facilitated in a grid formed by reducing the cross-linking density of the polymer because of its lateral flexible urethane branches.

The increased content of the modifier in the epoxy resin leads to higher impact resistance of coatings. This effect is probably associated with an increase in molecular mobility because of the introduction of flexible urethane units in the spatial structure of the polymer matrix, which contributes to the dissipation of mechanical energy input stroke [3].

It should be noted that the content of the modifier in the epoxy resin over 5 wt. per cent led to a marked gas release in the mixture and, consequently, to the defectiveness of the produced film.

TABLE 11.2 Physical and mechanical properties of the coatings modified by polyisocyanate

Temperature range	Modifier content (%)	Hardness (rel. u.)	Bending strength (mm)	Impact strength (cm)	Adhesion (score)
I range	0.0	0.95	20	30	1
	1.0	0.74	12	30	1

TABLE 11.2   *(Continued)*

Temperature range	Modifier content (%)	Hardness (rel. u.)	Bending strength (mm)	Impact strength (cm)	Adhesion (score)
	2.0	0.59	3	35	1
	3.0	0.49	1	45	1
	4.0	0.42	1	50	1
	5.0	0.37	1	50	1
II range	0.0	0.96	15	35	1
	1.0	0.89	3	45	1
	2.0	0.72	1	50	1
	3.0	0.60	1	50	1
	4.0	0.52	1	50	1
	5.0	0.48	1	50	1

It was also determined that the increase in the polyisocyanate to 5 wt. per cent causes significant deterioration of the protective characteristics of the coating because of their high water absorption, that can be explained by the increased porosity and lower density of the protective film, probably because of a significant plasticizing effect of the modifier. Owing to the fact that the steel constructions and devices are used not only indoors, but also outdoors, in the water environment often aggressively affecting the surface of the metal, it seemed reasonable to evaluate the water absorption of protective layers of the formed coatings from the developed film-forming epoxy compositions.

Water absorption was determined by assessing the sorption capacity of varnish to water (see Figure 11.1).

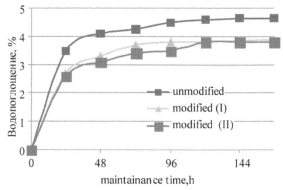

FIGURE 11.1   Water absorption of modified epoxy coatings.

It is known that epoxy film formers possess high chemical resistance to concentrated acids and alkalis, and are widely used in chemical industry, for this reason it is useful to study the effect of the modifier on the chemical resistance of the coatings.

The effect of the modifier on the protective properties of the coatings formed at different temperatures has been studied. Relative evaluation of protective characteristics after exposure of samples in hostile environments for a month, performed with the use of the method [4], is given in Table (11.3).

As seen from Table (11.3), the obtained composites exhibit improved protective properties with respect to water, 3 per cent NaCl, 10 per cent NaOH, and 25 per cent H_2SO_4. These coatings can be recommended for use in the chemical industry to protect tanks, reactors, and pipelines working in direct contact with hostile environments from corrosion.

TABLE 11.3 The results of testing coatings exposed in various environments for a month

Modifier content (%)	Relative value (%)				
	H_2O	3% NaCl	10% NaOH	25% H_2SO_4	Petroleum solvent C2
0	60	40	50	20	100
3 (I drying mode)	80	70	70	50	100
3 (II drying mode)	80	70	70	50	100

Quantitative testing of the modifier content of more than 5 wt. per cent of isocyanate in epoxy compositions was not conducted, for the films are defective with a large number of pores due to gas release; however, a visual tendency of films to turbidity was seen, which may indicate a significant increase in water absorption with increasing modifier content in the film-forming composition.

11.3 CONCLUSION

New film-forming compositions on the basis of epoxy resin ED-20 and industrially produced isocyanate were synthesized.

Protective coatings with improved adhesion, high impact resistance, moisture and water resistance, and high resistance to other hostile environments were obtained on their basis.

Optimal modifier content at which lacquer composition possesses the highest strain strength and protective characteristics has been determined.

It has been determined that the polyisocyanate modifier catalyzes the hardening process of epoxy oligomers and increases their resistance to corrosion.

KEYWORDS

- **Chemical modification**
- **Cross-linked polymer**
- **Epoxy materials**
- **Epoxy resin**
- **Film-forming composites**
- **Mechanical and chemical properties**
- **Polyurethane materials**

REFERENCES

1. Prokopchuk, N. R.; Chemistry and technology of film-forming materials. Prokopchuk, N. R.; Kruts'ko, E. T.; Minsk: BSTU; **2004**, 423.
2. Electrically insulating varnish for enameled wires: pat. 2111994 RF C09D5/25/ M. S. Fedoseyev [et.al.]; Institute of Technical Chemistry, the Urals. RAS. 95122362/04; published 27.05.98; **1998**, 6.
3. Zharin, D. E.; Polyisocyanate effect on physical and mechanical properties of epoxy composites. Zharin, D. E.; *Plastics*; **2002**, *7*, 38–41.
4. Karyakina, M. I.; Laboratory practical tutorial on testing paint-work materials and coatings. Karyakina, M. I.; M.: Chemistry; **1977**, 240.

CHAPTER 12

A RESEARCH NOTE ON THE MECHANISM OF DEVELOPMENT OF A MYOCARDIAL INFARCTION

A. A. VOLODKIN and G. E. ZAIKOV

CONTENTS

12.1 INTRODUCTION

It is known [1, 2] that at disintegration of myocardium cages, free radicals that can influence the mechanism of development of a heart attack are formed. Now the myocardial infarction considered as a consequence of sharp hydroxaemia because of damage of coronary vessels, and at formation of free radicals reaction with oxygen that can lead to the chain mechanism is possible. This direction in addition strengthens the deficiency of oxygen that often leads to a lethal outcome. It is known [3] that in reaction with oxygen peroxyradical radicals responsible for continuation of chain reaction are formed, and antioxidants inhibited this direction [4]. It is possible to assume that antioxidant application can render medical effect at a stage of development of myocardial infarction.

In this study as water-soluble and slowly toxic antioxidant is used 1-(carboxy)-1-(N-acetylamino)- 2 -(3', 5'-di-tert-butyl-4'-hydroxyphenyl)-propionate sodium (Anphen-sodium) possessing properties of radio- and cancer protectors [5, 6]. Diagnostic criterion of a heart attack of a myocardium, along with electrocardiogram methods, tests on "Troponin I" or "Troponin T". [7],

12.2 EXPERIMENTAL PART

1-(Carboxy)-1-(N-acetylamino)-2-(3', 5'-di-tert-butyl-4'-hydroxyphenyl)- propionate sodium (Anphen-sodium) received on a method [8].

Nuclear magnetic resonance spectrum ^{1}H (Д$_2$O, δ, m.): 1.36 (s, 18 H, ^{t}Bu); 1.96 (s., 3 H, COCH$_3$); 3.38 (s, 2H, CH$_2$); 4.70 (s, HOH);); 6.42–6.46 (s. уш., 1 H, OH); 6.95 (s, 2H, Ar); a nuclear magnetic resonance ^{13}C (Д$_2$O, δ, m.): 22.30 (CH$_3$); 29.43 (CH$_3$, ^{t}Bu); 33.64 (CH$_2$); 37.45 (C, ^{t}Bu); 125.94 (CH, Ar); 130.02 (C–CH$_2$, Ar); 139.18 (C–^{t}Bu), Ar); 150.80 (C–OH, Ar); 171.14 (C–N (O); 174.30 (C–OO). Let us dissolve in water and a physiological solution.

Studying of biological properties of a preparation "Anphen-sodium" with the use of epinephrine model of a heart attack of a myocardium (in updatings [9, 10]) spent under anesthetic rats of line Wistar (males, females) with weight of 200–220 g. Experimental animals have been divided into two groups on five rats in everyone.

First group—epinephrine introduction entered in a dose of 4 mg/kg.

Second group—after epinephrine introduction in (4 mg/kg), to rodents entered a preparation "Anphen-sodium" in a dose of 50 mg/kg.

For an estimation of efficiency of application of a preparation supervised following indicators:

- Survival rate of animals after epinephrine introduction;
- Changes on the electrocardiogram which has been removed at 4h after introduction by an animal of epinephrine. Cardiograms removed on electrocardiograph CardiMax FX-7102 in II standard assignment (II—left back and right an extremity lobby) in back position;
- Presence in blood of rats Troponin (TnI) in 4h after epinephrine introduction. Maintenance definition Troponin I carried out test device Troponin 1-Heck-1, testing on formation of the painted strip at interaction of antibodies to "TnI" from the sample of blood.

12.3   RESULTS

In case application of "Anphen-sodium" after epinephrine introduction prevented survival of experimental animals were in four of five experiments of testing rates. Before tests, animals behaved usually, deviations in their behavior, forage, and water consumption have not been noted. Electrocardiograms of all animals were in norm (Figure 12.1)

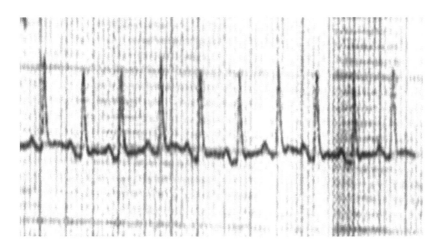

FIGURE 12.1   Electrocardiogram of a rat in the second standard assignment before tests.

In the first group, death rate within the first 48 h after adrenaline introduction was 100 per cent. Observed at opening of the fallen animals, pathoanatomical changes specified in a hypostasis of lungs, apparently, cardio-nature. Rats of second group at 24 h after introduction of epinephrine and "Anphen-sodium" remained live though in the first 2 days showed strong block; at two rats of this group it was marked breath by a mouth, spasms. At 4h after introduction of epinephrine of the electrocardiogram has shown.

First Group: The registration of lifting of segment "ST" against sharp reduction of height of tooth "R" and delays of a warm rhythm. The height of tooth "R" decreases at the expense of formation of a zone of damage to a myocardium (Figure 12.2).

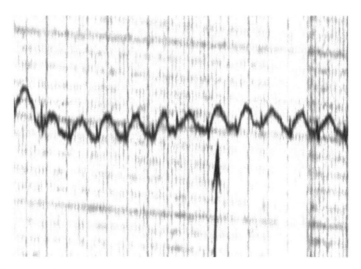

FIGURE 12.2   Electrocardiogram of a rat in the second standard assignment from the first group.

At 48 h after introduction of epinephrine, two rats were decapitated. In blood "Troponin I"

Second Group: Electrocardiogram registers lifting of segment "ST" above an isoline and its merge to positive tooth «T» (Figure 12.3).

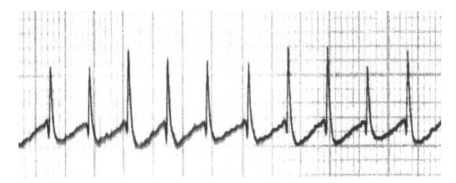

FIGURE 12.3   Electrocardiogram of a rat in the second standard assignment from the second group.

For 20 days four rats from five experiments continued to live with signs of languid behavior.

12.4   DISCUSSION

Influence of antioxidants on processes in a myocardium in conditions hypoxaemia was marked earlier [11]; however their role in heart attack development was not discussed. Taking into account radical processes at necrosis of ischemia cardiomiotsity, it is necessary to consider process of interaction of oxygen with radicals that predetermines the possibility of development of a heart attack of a myocardium on the chain mechanism. Participation in this process of an antioxidant slows down speed of chain reaction that should lead to more limited area of a heart attack of a myocardium with preservation of viability of an organism. Structure "Anphensodium" provides high antioxidizing parameters and ability to take root in lipid a cover erythrocytus that gives to the given preparation unique properties. Results confirm that in an initial stage of development of a heart attack of a myocardium in experiences invivo on rats "Anphen-sodium" renders medical result. On the electrocardiogram of an animal after introduction of epinephrine and further it is consecutive physiological a solution "Anphen-sodium" lifting of segment «ST» (Figure 12.3) is observed that can specify in an initial stage of development of a myocardium infarction. At the same time the height of tooth «R» does not change, whereas in experiment from fist group (Figure 12.2) tooth «R» is hardly noticeable.

In comparison with control the survival rate makes four animals from five experimental whereas in control 100 per cent death rate are marked. This result can be interpreted as consequence of inhibition of development of a heart attack of a myocardium under the influence of an antioxidant and partially radical nature of this pathological process. At a stage before infarct conditions as a result hypoxemia there is an accumulation of radicals to some critical concentration and further the chain mechanism is started interactions of radicals with oxygen (a stage of development of a heart attack of a myocardium). The mechanism of similar processes is described [4] which proceeds under the Scheme (12.1):

$$RH \longrightarrow \overset{\bullet}{R} \qquad (1)$$

$$\overset{\bullet}{R} + O_2 \longrightarrow R\overset{\bullet}{O}_2 \qquad (2)$$

$$R\overset{\bullet}{O}_2 + RH \longrightarrow RO_2H + \overset{\bullet}{R} \qquad (3)$$

$$R\overset{\bullet}{O}_2 + InH \longrightarrow RO_2H + \overset{\bullet}{In} \qquad (4)$$

Formation of radicals from cardiomiotsity (R) at a myocardial infarction is known [2, 3] (reaction type1). Reactions type (12.2) and (12.3) specify in the oxygen expense on the chain mechanism that can correspond to a stage of development of a heart attack on the chain mechanism. Antioxidants (InH) on reaction (12.4) stop this process.

12.5 CONCLUSIONS

Qualitative results on the basis of electrocardiograms of experimental animals confirm efficiency of a preparation "Anphen-sodium" in possibility of prevention of development of a heart attack with a lethal outcome.

KEYWORDS

- 1-(carboxy)-1—(N-acetylamino) -2—(3', 5'-di-tert-butyl-4'-hydroxyphenyl)—propionate sodium
- Epinephrine
- Myocardial infarction

REFERENCE

1. Berdyshev, D. V.; Glazunov, V. P.; and Novikov, V. L.; Izv. The Russian Academy of Sciences. Ser.Khim.; 2007, 400, (Russian).
2. Mishchenko, N. P.; Fedoreev, S. A.; and Bagirova, V. L.; Fharm. Chem. J. 2003, 37, 49, (Engl. Transl.)
3. Emanuel, N. M.; Denisov, E. T.; Majzus, Z. K.; Chain Reactions in a Liquid Phase. M.: Science; 1965, (Russian).
4. Emanuel, N. M.; Ljaskovskaja, J. N.; Braking of Processes of Oxidation of Fats. M.: Pishchepromizdat; 1961, (Russian).
5. Volodkin, A. A.; Burlakova, E. B.; Zaikov, G. E.; Gaintseva, V. D.;Lomakin, S. M.; and Shevtsov, S. A.; Vestn. Kasan. Technolog. University; 2012, 18(4), 99, (Russian).
6. Volodkin, A. A.; Erokhin, V. N.; Burlakova, E. B.; Zaikov, G. E.; Lomakin, S. M.; J. Chem. Phys.2013, 32, 66–72, (Russian).
7. Trifonov, I. R.; Kardiologija, Z. 2001, 11, 93–95, (Russian).
8. Volodkin, A. A; Lomakin, S. M; Zaikov, G. E.; Evteeva, N. M; Izv. The Russian Academy of Sciences, Ser. Khim.; 2009, 900, (Russian).
9. Chuvaev, I. V.; Materials of the All-Russia Congress of Veterinary Pharmacologists. 2009, 28–31, (Russian).
10. Nepomnjashchih, L. M.; Alternative Insufficiency of Muscular Cages of Heart at Metabolic and Ischemic Damages. M.; 1998, 56, (Russian).
11. Lakomkin, V. L.; Abramov, A. A.; Kapelko, V. I.; and Kardiologija, Z.; 2011, 11, 60–64, (Russian).

CHAPTER 13

A COMMENTARY NOTE ON STRUCTURAL TRANSFORMATIONS OF PIRACETAM UNDER LEAD ACETATE INFLUENCE

O. V. KARPUKHINA, K. Z. GUMARGALIEVA, S. B. BOKIEVA, and A. N. INOZEMTSEV

CONTENTS

13.1　INTRODUCTIONS

Piracetam (PIR), a standard representative of class of psychotropic medicals—nootrops [1], has been extensively used recently both in clinic practice and experimental studies. Piracetam features diverse functional properties and ambiguous effect on learning capability and memory, even if introduced in small doses in various test series [2, 3].

It is natural to attribute this ambiguity to some unaccounted physicochemical factors. Among them we can mention different contents in the medical of conformers—polymorphic crystalline structures, which can give rise to various permolecular structures in PIR solutions [4].

The PIR effect diversity can also be affected by physicochemical characteristics of water, in this respect, of great interest are salts of heavy metals, for example, lead acetate, contained in water, including drinking water not purified in biological filters, in appreciable quantities. If this supposition is true, in regions with a high metal content in water one can expect unusual effects of PIR on superior neural human activity because of its extensive application in clinic practice.

The aforesaid stimulated us, on the one hand, to study the effect of PIR solved in water-containing lead acetate on leaning and memory of rats and, on the other to ascertain changes in the PIR structure brought about by lead acetate additives.

We used 45 white strain-free male rats 180–200 g in weight divided into four groups: we introduced interperitoneally PIR in amount of 300 mg/kg in rats of group I half an hour before each test, lead acetate solution ($10-7$ mol/l) in rats of group II 4.5 h before test, lead acetate solution and PIR in rats of group III, and physiological solution in rats of group IV (reference group). Then, for 5 days we developed (25 trials per day) an "avoidance" reflex in a shuttle chamber. Tests were carried out as follows: we switched on an irritant (800 Hz sound) and then in 10 s, electric current. Running of animals to another half of the chamber switched off both stimuli. The pause between signals was 30–60 s.

To study the structural changes, we prepared 2 per cent (weight) PIR solutions in distilled water (solution I) and in 10^{-7} mol/l lead acetate solution in distilled water (solution II). Changes in the state of the solutions were analyzed by liquid chromatography.

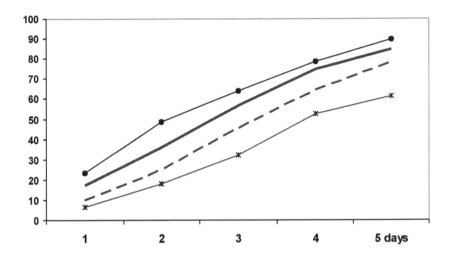

FIGURE 13.1   Dynamics (%) of avoidance of electric shock by rats injected with aqueous solutions if piracetam, lead acetate, and their mixture: —●—PIR ($p < 0.001$), - - - - - Pb acetate, —— reference, —ж— PIR + Pb.

Chromatograms are taken on a Beckman chromatograph with reversed ODS phase, isocratic elution (1 l/min), methanol and water (15:85 volume ratio) as eluents, and detection at a wave length equal to 220 nm. Calorimetric measurements were conducted on a Mettler—3301 differential scanning calorimeter (DSC) at a heating rate of 1°/min rate.

The experimental data displayed in Figure (13.1) suggest that neither PIR nor lead acetate introduced separately suppress learning. Moreover, the learning curve for animals injected with PIR always goes above the reference, although this surplus attains a statistically significant level only on the third day ($p = 0.01$). However, the combined effect of these solutions suppresses learning, which shows up in a statistically significant reduction of avoidance reactions in the second test. These agents exerted a similar effect on the locomotion activity in the form of intersignal reactions, that is, individually they did not suppress this activity, but together, they did.

A representative chromatographic trace of a PIR solution is show in Figure (13.2).

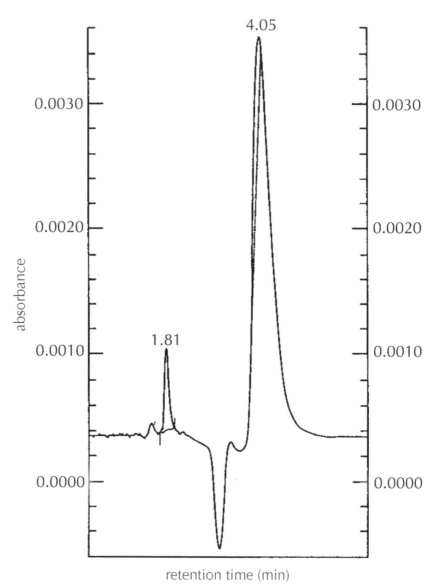

FIGURE 13.2   Chromatographic record of a newly prepared aqueous solution of Piracetam.

Lead acetate additives and solution storage do not change the signal retention time τ affecting solely the areas under the appropriate curves. To assign the chromatographic peaks, we ascertained the dependence of phase transition heats and chromatographic records of aqueous solution of PIR on the type of its thermal pretreating (Table 13.1).

The measurements show that the medicals used features two melting points (400 and 420 K). This is supported by the data on the polymorphic structure of crystalline PIR [4].

Changes in the phase transition heart entail variations of the UV signal in chromatographic analysis of aqueous solution of PIR. Changes in the heat at 400 and 420 K are correlated with amplitude of signals $\tau = 1.81$ min and $\tau = 4.05$ min, respectively. It means that piracetam is an associate of different crystalline structures. Inasmuch as the temperatures at which the DSC curve exhibits an endothermic peak are independent of thermal pretreatment modes and the PIR sample weight does not change, a decrease in the total phase transition heart (Table 13.1) at a constant specific heat implies that the amount of crystalline PIR in samples reduces as a result of thermal treatment.

This also signifies that along with the crystalline phase, solid PIR also contains a noncrystalline (amorphous) phase. When one of the crystalline structures is completely transformed into the amorphous phase (Table 13.1, line 2), the appropriate endothermic peak in the DSC record disappears; however, the appropriate signal in the UV range partially remains. This suggests that solution of the amorphous PIR phase also absorbs in the UV range, though with a lower (about 30 time as low) extinction coefficient. The contribution of amorphous PIR to the absorption in the UV range is the major reason why a relative reduction of the UV signal upon one thermal treatment or another slightly differs (by 2–10 per cent) from the reduction of melting heats under the same conditions.

Acetamide dimers present, as Figure (13.3) shows, solely in the crystalline state with a melting point of about 400 K can serve as these structure elements.

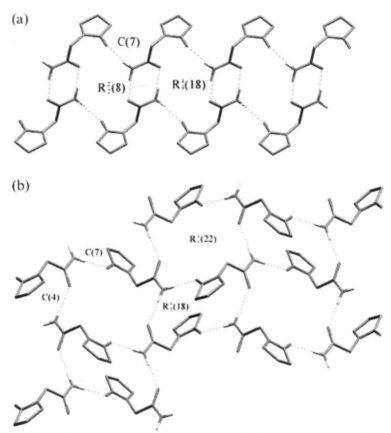

FIGURE 13.3 The molecular packing arrangements in the crystal structures of piracetam [Fabbiani, Allan, Parsons and Pulham, 2005]. (a) for the melting point of about 420 K. Acetamide dimers are framed and (b) is the structure pertaining to the polymorphic phase with a melting point of 400 K.

The results of chromatographic measurements of changes occurring in aqueous solutions at various times lapsed since solution preparation are summarized in Table (13.2). Both in solutions with lead acetate additives and proper PIR the peak areas at $\tau = 1.81$ min diminish, although at different rates, to values close to the appropriate values recorded in newly prepared solutions of amorphous PIR. The peak at $\tau = 4.05$ min in solution without lead acetate is stable, while the peak area of solutions without lead acetate slightly rises with time.

A decrease in the signal at $\tau = 1.81$ min signifies disordering of the initial structure. The different behaviors of the peaks at $\tau = 4.05$ min is associated with the fact that PIR even in its amorphous state retains some structure elements in the absence of lead acetate. These residual structure elements preclude formation of stable associates with structures inherent in $\tau = 4.05$ min.

Lead acetate present in solution II and capable of producing strong complexes with amide groups destroys acetamide dimers, thereby facilitating transition from one structure to another. The above-mentioned "suppression" effect is presumably associated with a significant increase in the content of the biologically active dissociation product of the aforesaid dimer in the solution.

TABLE 13.1   The effect of thermal treatment mode on phase transition heats of solid Piracetam and chromatographic records of newly prepared aqueous solutions

N/N	The sort of heat treatment	Phase transition heat J/g, at temperature (K)		Chromatogram peaks area (relative units) at the output time (min)	
		400	**420**	**1.81**	**4.05**
1	None	24.4	182.2	6.52	382.5
2	Heating up to 475 K and cooling up to 300 K	0	170.5	0.189	351.0
3	Heating up to 475 K, cooling up to 300 K and 2 weeks standing	9.2	172.0	2.98	355.0
4	Heating up to 525 K and cooling up to 300 K	13.4	110.0	4.06	240.8

TABLE 13.2 The dependence of changing of chromatogram peaks upon the time from the moment of piracetam solutions preparations II and I

Time from the moment of solution preparation (min)	Area of peak (relative)units for two output times (min)			
	Solution I		Solution II	
	1.81	4.05	1.81	4.05
0	6.52	382.5	6.45	377.7
15	4.12	382.5	0.72	378.1
33	1.05	382.5	0.29	382.27
66	0.41	382.5	0.21	389.99
101	0.22	382.5	0.12	389.99
143	0.22	382.5	0.12	395.13

KEYWORDS

- **Piracetam**
- **Polymorphism**
- **Structural Transformations**
- **Lead acetate**

REFERENCE

1. Giurgea, C. E.; The nootropic concept and its prospective implications. *Drug. Dev. Res.* **1982**, *2*, 441–446.
2. Burov, Yu.; Inozemtsev, A.; Pragina, L.; Litvinova, S.; Karpukhina, O.; Tushmalova, N.; *Bull. Exper. Biology and Medicine.* **1993**, *115(2)*, 150–152.
3. Inozemtsev, A.; Kapitsa, I.; Garibova, T.; Bokieva, S.; Voronina, T.; *Mos. Un. Biological Sciences Bulletin.* **2004**, *3*, 24–26.
4. Fabbiani, F. P. A.; Allan, D. R.; Parsons, S. and Pulham, C. R.; *Cryst. Eng. Comm.* **2005**, *7*, 179–186.

CHAPTER 14

A RESEARCH NOTE ON TRANSPORT PROPERTIES OF FILMS ON A BASIS CHITOSAN AND MEDICINAL SUBSTANCE

A. S. SHURSHINA, E. I. KULISH, and S. V. KOLESOV

CONTENTS

14.1 INTRODUCTION

Systems with controlled transport of medicines are the extremely demanded [1, 2]. Research of different processes on diffusion of water and medicinal substance in polymer films and opportunities of control of release of medicines became the purpose of this works. As a matrix for the immobilization of drugs used in naturally occurring polysaccharide chitosan, which has a number of valuable properties: non-toxicity, biocompatibility and high physiological activity [3], as well as a drug used aminoglycoside antibiotic series—amikacin, actively applied in the treatment of pyogenic infections of the skin and soft tissue [4]

14.2 EXPERIMENTALS

The main object of investigation was a chitosan (ChT) specimen produced by the company "Bioprogress" (Russia) and obtained by acetic deacetylation of crab chitin (degree of deacetylation ~84%) with M_{sd} = 3,34,000. As the medicinal substance (MS) used an antibiotic amikacin (AM)—quadribasic aminoglycoside, used in the form of salts—sulfate (AMS) and chloride (AMCh). Chemical formulas of objects of research and their symbols used in the text are given in Table (14.1).

TABLE 14.1 Formulas research objects and symbols used in the reaction schemes

Formula object of study	Symbol
Chitosan acetate monomer unit CH_2OH O OH $NH_3^+CH_3COO^-$	$\sim\!\sim\!\sim NH_3^+CH_3COO^-$

amikacin sulfate	
amikacin chloride	

ChT films were obtained by means of casting of the polymer solution in 1 percent acetic acid onto the glass surface with the formation of chitosan acetate (ChTA). Aqueous antibiotic solution was added to the ChT solution immediately before film formation. The content of the medicinal preparation in the films was 0.01, 0.05, and 0.1 mol/mol ChT. The film thickness in all experiments was maintained to be constant and equal to 100 μ. To study the release kinetics of MS, the sample was placed in a cell with distilled water. Stand out in the aqueous phase AM recorded spectrophotometrically at a wavelength of 267 nm, corresponding to the maximum absorption in the UV spectrum of MS. The quantity of AM released from the film at time t (G_s) was estimated from the calibration curve. The establishment of a constant concentration in the solution of MS G_∞ is the time to equilibrium. MS mass fraction α is available for diffusion, assessed as the quantity of films released from the antibiotic to its total amount entered in the film.

Studying of interaction MS with ChT was carried out by the methods of IR- and UV-spectroscopy. IR-spectrums of samples wrote down on spectrometer "Shimadzu" (the tablets KBr, films) in the field of 700–3,600 cm^{-1}. The UV spectrums of all samples removed in quartz ditches thick of 1 cm concerning water on spectrophotometer "Specord M-40" in the field of 220–350 nm.

With the aim of determining the amount of medicinal preparation held by the polymer matrix β, the synthesis of adducts of the ChT-antibiotic interaction was carried out in acetic acid solution. The synthesized adducts were isolated by double re-precipitation of the reaction solution in NaOH solution, followed by washing of precipitated complex residue with isopropyl alcohol. Then, the residue was dried in vacuum up to constant mass. The amount of preparation strongly held by chitosan matrix was determined according to the data of the element analysis on the analyzer EUKOEA—3,000 and UF-spectrophotometrically.

The relative amount of water m_t absorbed by a film sample of ChT, determined by an exiccator method, maintaining film samples in vapors of water before saturation, and calculated on a formula: $m_t = (\Delta m)/m_0$, where m_0 is the initial mass of ChT in a film, Δm is the weight of the absorbed film of water by the time t.

Isothermal annealing of film samples carried out at a temperature 120°C during fixed time. Structure of a surface of films estimated by method of laser scanning microscopy on device LSM-5-Exciter (Carl Zeiss, Germany).

14.3 RESULTS AND DISCUSSION

It is well known that the release of MS from polymer systems proceeds as diffusive process [5–7]. However, a necessary condition of diffusive transport of MS from a polymer matrix is its swelling in water, that is, effective diffusion of water in a polymer matrix by diffusing in a polymer matrix. The water molecules possessing it is considerably bigger mobility in comparison with high-molecular substance, and it penetrates in a polymer material, separating apart chains and increasing the free volume of a sample. The main mechanisms in water transport in polymer films are simple diffusion and the relaxation phenomena in swelling polymer. If transfer is caused mainly the processes mentioned, the kinetics of swelling of a film is described by the following equation [8]:

$$m_t/m_\infty = kt^n \tag{14.1a}$$

where m_∞ is the relative amount of water in equilibrium swelling film sample, is the constant connected with parameters of interaction polymer–diffuse substance, and n is an indicator, characterizing the mechanism of transfer of substance. If transport of substance is carried out on the diffusive mechanism, the indicator of n has to be close to 0.5. If transfer of substance is limited by the relaxation phenomena, then $n > 0.5$.

The parameter n determined for a film of pure ChT is equal to 0.63 (i.e., >0.5) that is characteristic for the polymers, being lower than that of vitrification temperature [9]. This fact is connected with slowness of relaxation processes in glassy polymers. The main values of equilibrium sorption of water and indicator n defined for film samples passed isothermal annealing (a relaxation of non-equilibrium conformations of chains with reduction of free volume) are presented in Table (14.2).

TABLE 14.2 Parameters of swelling of the chitosan films in water vapor

Composition of the film	The concentration of MS in the film (mol/mol ChT)	Annealing (time, min)	$D_s^a * 10^{11}$ (cm²/sec)	$D_s^6 * 10^{11}$ (cm²/sec)	n	Q_∞ (g/g ChT)
ChT		15	37.2	37.0	0.50	2.50
		30	36.3	36.0	0.48	2.48
		60	15.3	14.1	0.44	2.47
		120	12.2	13.0	0.43	2.46
ChT-AMCh	0,01	30	8.5	4.7	0.42	1.84
		60	7.0	4.5	0.39	1.56
		120	5.8	4.2	0.37	1.42
	0.05	30	5.3	3.2	0.34	1.85
	0.1	30	4.4	3.1	0.32	1.48

TABLE 14.2 *(Continued)*

Composition of the film	The concentration of MS in the film (mol/mol ChT)	Annealing (time, min)	$D_s^a *10^{11}$ (cm²/sec)	$D_s^6 *10^{11}$ (cm²/sec)	n	Q_∞ (g/g ChT)
ChT-AMS	0.01	30	6.9	2.9	0.34	1.58
		60	4.0	2.8	0.27	1.46
		120	2.6	2.3	0.25	1.31
	0.05	30	6.1	2.2	0.30	1.66
	0.1	30	2.7	1.9	0.27	1.07

Apparently from the data given Table (14.2), carrying out isothermal annealing leads to the values of an indicator n decrease. Thus, if annealing was carried out during small time (15–30 min), the value *n* determined for pure ChT is close to 0.5. It indicates that transfer of water is limited by diffusion, and it is evidence that ChT in heat films is in conformational relaxed condition. In process of increase time of heating until 60–120 min, values of an indicator *n* continue to decrease that, most likely reflects process of further restructure of the polymer matrix, occurring in the course of film heating. The processes of isothermal annealing of ChT at temperatures $\geq 100°C$ are accompanied by course of a number of chemical transformations, was repeatedly noted in literature [10, 11]. In particular, it is revealed that besides acylation reaction, there is the partial destruction of polymer increasing the maintenance of terminal aldehyde groups, which reacting with amino groups, sew ChT macromolecules at the expense of formation of azomethine connections. In [12] the fact of cross-linking in the HTZ during isothermal annealing was confirmed by the study of the spin-lattice relaxation. In the values of equilibrium sorption isothermal annealing, however, actually no clue, probably owing to the low density of cross-links.

A similar result—reduction indicator n is achieved when incorporated into a polymer matrix MS. As per the data in Table (14.2), the larger the MS entered into the film, the slower and less absorb water ChT. During isothermal annealing, medicinal films effect get enhanced. Such deviations from the laws of simple diffusion (Fick's law) and others researches have been observed, explaining their strong interaction polymer with MS [13].

In the aqueous environment from ChT film with antibiotic toward the water flow moving to volume of the chitosan, from a polymeric film LV stream is directed to water.

In Figure (14.1) typical experimental curves of an exit of AM from chitosan films with different contents of MS are presented. All the kinetic curves are located on obviously expressed limit corresponding to an equilibrium exit of MS (G).

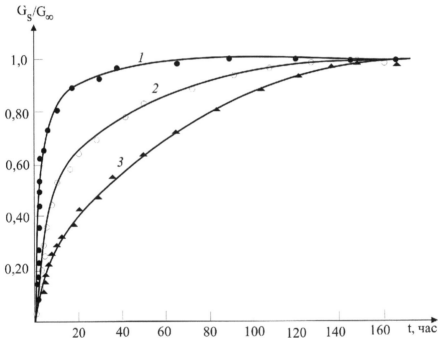

FIGURE 14.1 Kinetic curves of the release of the MS from film systems ChT-AMCh with the molar ratio of (1) 1:0.01; (2) 1:0.05; and (3) 1:0.1. Isothermal annealing time—30 min.

Mathematical description of the desorption of low-molecular-weight component from the plate (film) with a constant diffusion coefficient is in detail considered in [14], where the original differential equation adopted formulation of second Fick's law:

$$\frac{\partial c_s}{\partial t} = D_s \frac{\partial^2 c_s}{\partial x^2}$$

(14.1b)

The solution of this equation disintegrates into two cases: one for big ($G_s/G_\infty > 0.5$) and the other for small ($G_s/G_\infty \leq 0.5$) experiment times:

On condition $G_s/G_\infty \leq 0.5$ $G_s/G_\infty = [16D_s t/\pi L^2]^{0.5}$ (14.2a)

On condition $G_s/G_\infty > 0.5$ $G_s/G_\infty = 1 - [8/\pi^2 \exp(-\pi^2 D_s t/L^2)]$ (14.2b)

where $G_s(t)$—concentration of the desorbed substance at the time t and G_∞—Gs value at $t \to \infty$, and L—thickness of the film sample.

In case of transfer MS with constant diffusion coefficient, the values calculated as on initial (condition $G_s/G_\infty \leq 0.5$), and at the final stage of diffusion coefficient (condition $G_s/G_\infty > 0.5$), must be equal. The equality $D_s^a = D_s^b$ indicates the absence of any complications in the diffusion system polymer—low-molecular substance [15]. However, apparently from the data presented in Table (14.3), for all analyzed cases, the value of diffusion coefficients calculated at an initial and final stage of diffusion do not coincide.

TABLE 14.3   Desorption parameters AMS and AMCh in the films based on ChT

Composition of the film	The concentration of MS in the film (mol/mol ChT)	Annealing time (min)	$D_s^a \ast 10^{11}$ (cm²/sec)	$D_s^6 \ast 10^{11}$ (cm²/sec)	n	α
ChT-AMCh	1:0.01	30	81.1	3.3	0.36	0.95
		60	76.9	3.0	0.33	0.92
		120	37.0	2.9	0.25	0.88
	1:0.05	30	27.2	2.5	0.29	0.91
	1:0.1	30	25.9	2.2	0.21	0.83
ChT-AMS	1:0.01	30	24.6	1.9	0.30	0.94
		60	23.0	1.8	0.22	0.90
		120	21.1	1.4	0.20	0.85
	1:0.05	30	24.0	1.8	0.18	0.80
	1:0.1	30	23.3	1.5	0.16	0.70

Note that in the process of water sorption, similar regularities (Table 14.2) were observed. This indicates a deviation of the diffusion of classical type and suggests the so-called pseudo-normal mechanism of diffusion of MS from a chitosan matrix.

About pseudonormal-type diffusion MS also shows kinetic curves, constructed in coordinates $G_s/G_\infty - t^{1/2}$ (Figure 14.2). In the case of simple diffusion, the dependence of the release of the MS from film samples in coordinates $G_s/G_\infty - t^{1/2}$ it would have to be straightened at all times of experiments. However, as can be seen from Figure (14.2), a linear plot is observed only in the region $G_s/G_\infty < 0.5$, after which the rate of release of the antibiotic significantly decreases.

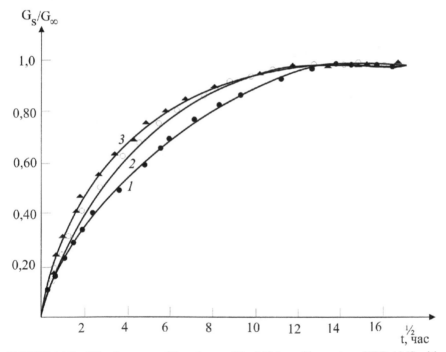

FIGURE 14.2   Kinetic curves of the release of the MS from film systems HTZ-AMS with a molar concentration of (1) 0.01; (2) 0.05; and (3) 0.1. Isothermal annealing time—30 min.

At diffusion MS from films, also as well as in case of sorption by films ChT—AM of water vapor, have anomalously low values of the parameter n, estimated in this case from the slope in the coordinates $\lg(G_s/G_\infty)$—$\lg t$.

Increase the concentration of MS and time of isothermal annealing, as well as in the process of sorption of water vapor films, accompanied by an additional decrease in the parameter n. Symbiotically, index n also changes magnitude α. Moreover, the relationship between the parameters of water sorption films ChT-AM (values of diffusion coefficient, an indicator n and values of equilibrium sorption Q) and the corresponding parameters of the diffusion of MS from the polymer film on condition $G_s/G_\infty \leq 0.5$ is shown.

All known types of anomalies of diffusion, can be described within relaxation model [16]. Unlike the Fikovsky diffusion assuming that instant establishment and changes in the surface concentration of sorbate, the relaxation model assumes change in the concentration in the surface layer on the first-order equation [17]. One of the main reasons causing change of boundary conditions is called nonequilibrium of the structured morphological organization of a polymer matrix [16]. A possible cause of this could be the interaction between the polymer and LV.

Abnormally low values n in work [18] were explained with presence tightly linked structural of amorphous-crystalline matrix. In this case, is believed, the effective diffusion coefficient in process of penetration into the volume of a sample can be reduced due to steric constraints that force diffusing by pass crystalline regions and diffuse to the amorphous mass of high-density cross-linking. As ChT belongs to the amorphous-crystalline polymers, it would be possible to explain low values n observed in our case similarly. However, according to Table (14.2) in the case of films of individual ChT such anomalies are not observed. Thus, the effect of substantially reducing n is associated with the interaction between HTZ and LV.

In the volume of the polymer matrix MS can be in different states. Part of the drug may be linked to the macromolecular chain through any chemical bonds and on the other hand, some of it may be in the free volume in the form of physical filler. In the latter case, it can be cause a certain structural organization of the polymer matrix. That antibiotic AM influences on the structure and morphology of the films ChT, which indicate data of the laser scanning microscopy. As seen from the electron microscope images of Figure (14.3), the initial films ChT visible surface strong interference caused by its heterogeneity. With the introduction of the films AM, the interference surface is significantly reduced.

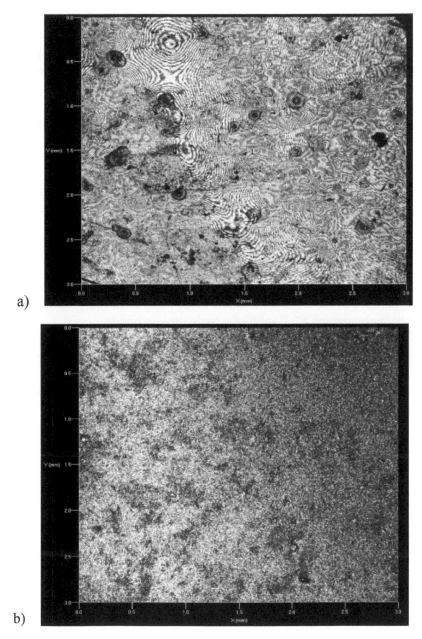

a)

b)

FIGURE 14.3 Micrograph of the surface (in contact with air) (a) film individual ChT and (b) film ChT-AMS.

IR-and UV-spectroscopy data indicate to taking place interaction between AM and ChT. It may be noted and for example, a significant change in the ratio between the intensity of the bands ChT corresponding hydroxyl and nitrogen-containing groups, before and after the interaction with the antibiotic (Table 14.4).

TABLE 14.4　The value of the intensity ratio of the absorption bands of some of the data of the IR-spectra:

Sample	I_{1640}/I_{2900}	I_{1590}/I_{2900}	I_{1458}/I_{2900}
ChT	0.57	0.51	0.72
AM	0.61	0.62	0.7
ChT-AM complex derived from 1% acetic acid	0.48	0.53	0.59

The binding energy in the adduct reaction ChT-antibiotic was estimated by the shift in the UV spectra of the order of 10 kJ/mol, which allows to tell about connection ChT-antibiotic by hydrogen bonds.

Thus, AM may interact with ChT by forming hydrogen bonds. However, the interpretation of the data on the diffusion is much more important than AM can form chains linking ChT by salts formation.

Exchange interactions between ChT and AMS may occur under the scheme:

Due to dibasic sulfuric acid, it is possible to suggest the formation of two types of salts, providing stapling ChT macromolecules with the loss of its solubility. Firstly, the water-insoluble "double" salt—sulfate ChT-AMS and secondly, the salt mixture—insoluble in water ChT sulfate and soluble AM acetate.

If the AM takes in the form of chloride, an exchange reaction between ChT acetate and AM chloride reduces the formation of dissociated soluble salts. Accordingly, the reaction product in this case will consist of the H-complex ChT-AM.

If data on a share of antibiotic related to polymer adducts (β), obtained in solutions of acetic acid, are presented in Table (14.5).

TABLE 14.5 Mass fraction of the antibiotic β, defined in reaction of adducts obtained from 1 per cent acetic acid

Used antibiotic	The concentration of MS in the film (mol/mol ChT)	β
AMS	1.00	0.72
	0.10	0.33
	0.05	0.21
	0.01	0.07
AMCh	1.00	0.37
	0.10	0.20
	0.05	0.08
	0.01	0.04

As seen in Table (14.5), from the fact that AMC is able to "sew" chitosan chain is significantly more closely associated with macromolecules MS than for AMX.

Formation of chemical compounds of MS with ChT is probably the reason for the observed anomalies—reducing the rate of release of MS from film caused by simple diffusion, as well as the reduction of the share allocated to the drug (α). Indeed, the proportion of MS found in reaction of adducts correlates with the share of antibiotic is not capable of participating in the diffusion process, and with the index n, reflecting the diffusion mechanism (Figure 14.4).

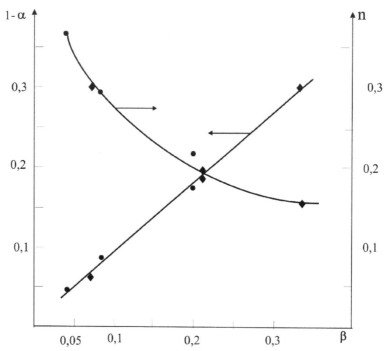

FIGURE 14.4   The fraction of amikacin, not involved in the diffusion of amikacin share determined in the adduct of reaction. ″—system ChT-AMS; ·—ChT-AMCh.

Thus, the structural changes in the polymer matrix, including as a result of its chemical modification of the interaction with the drug substance, cause deviations regularities of transport MS of chitosan films from classic Fikovskogo mechanism. Mild chemical modification, for example by cross-linking macromolecules salt formation, not affecting the chemical structure of the drug, is a possible area of control of the transport properties of medicinal chitosan films.

KEYWORDS

- **Chitosan**
- **Diffusion**
- **Medicinal substance**
- **Sorption**

REFERENCES

1. Shtilman, M. I.; Polymers of Medicobiological Appointment. M.: Akademkniga; **2006,** 58 p.
2. Plate, N. A.; and Vasilev, A. E.; Physiologically Active Polymers. M.: Chemistry; **1986,** 152 p.
3. Skryabin, K. G.; Vihoreva, G. A.; and Varlamov, V. P.; Chitin and Chitosan. Preparation, Properties and Application. M.: Science; **2002,** 365 p.
4. Mashkovskii, M. D.; Pharmaceuticals. Kharkov: Torsing; **1997,** *2,* 278 p.
5. Ainaoui, A.; and Verganaud, J. M.; *Comput. Theor. Polym. Sci.* **2000,** *10(2),* 383.
6. Kwon, J.-H.; Wuethrich, T.; Mayer, P.; and Escher, B. I.; *Chemosphere.* **2009,** *76,* 83 p.
7. Martinelli, A.; D Ilario, L.; Francolini, I.; and Piozzi, A.; *Int. J. Pharm.* **2011,** *407(1–2),* 197.
8. Hall, P. J.; and Thomas, K. M.; Marsh Fuel. **1992,** *71(11),* 1271 p.
9. Chalyih, A. E.; Diffusion in Polymer Systems. M.: Chemistry; **1987,** 136 p.
10. Zotkin, M. A.; Vihoreva, G. A.; and Ageev, E. P.; et al. Engineering Chemistry. **2004,** *9,* 15 p.
11. Ageev, E. P.; Vihoreva G. A.; and Zotkin, M. A.; et al. High-Molecular Compounds. **2004,** *46(12),* 2035.
12. Smotrina, T. V.; and Butlerovsky Messages. **2012,** *29(2),* 98–101.
13. Singh, B.; and Chauhan, N.; Acta Biomaterialia. **2008,** *4(1),* 1244.
14. Crank, J.; The Mathematics of Diffusion. Oxford: Clarendon Press; **1975,** 46 p.
15. Ukhatskaya, E. V.; Kurkov, S. V.; and Matthews, S. E.; et al. *Int. J. Pharm.* **2010,** *402(1–2),* 10.
16. Malkin, A. Ya.; and Chalyih, A. E.; Diffusion and viscosity of polymers. Methods of Measurement. M.: Chemistry; **1979,** 304 p.
17. Pomerancev, A. L.; Methods of non-linear regression analysis to model the kinetics of chemical and physical processes. In: Dissertation of the Doctor of Physical and Mathematical Sciences. M.: MGU; **2003.**
18. Kuznecov, P. N.; Kuznecova, L. I.; and Kolesnikova, S. M.; Chemistry in interests of a sustainable development. **2010,** *18,* 283–298.

CHAPTER 15

A RESEARCH NOTE ON APPLICATION OF POLYMERS CONTAINING SILICON NANOPARTICLES AS EFFECTIVE UV PROTECTORS

A. A. OLKHOV, M. A. GOLDSHTRAKH, and G. E. ZAIKOV

CONTENTS

15.1   INTRODUCTION

In recent years, considerable efforts have been devoted for search of new functional nanocomposite materials with unique properties that are lacking in their traditional analogs. Control of these properties is an important fundamental problem. The use of nanocrystals as one of the elements of a polymer composite opens up new possibilities for targeted modification of its optical properties because of a strong dependence of the electronic structure of nanocrystals on their sizes and geometric shapes. An increase in the number of nanocrystals in the bulk of composites is expected to enhance long-range correlation effects on their properties. Among the known nanocrystals, nanocrystalline silicon (nc-Si) attracts high attention due to its extraordinary optoelectronic properties and manifestation of quantum size effects. Therefore, it is widely used for designing new generation functional materials for nanoelectronics and information technologies. The use of nc-Si in polymer composites calls for a knowledge of the processes of its interaction with polymeric media. Solid nanoparticles can be combined into aggregates (clusters), and, when the percolation threshold is achieved, a continuous cluster is formed.

An orderly arrangement of interacting nanocrystals in a long-range potential minimum leads to the formation of periodic structures. Because of the well-developed interface, an important role in such systems belongs to adsorption processes, which are determined by the structure of the nanocrystal surface. In a polymer medium, nanocrystals are surrounded by an adsorption layer consisting of polymer, which may change the electronic properties of the nanocrystals. The structure of the adsorption layer has an effect on the processes of self-organization of solid-phase particles, as well as on the size, shape, and optical properties of resulting aggregates. According to data obtained for metallic [1] and semiconducting [2] clusters, aggregation and adsorption in three-phase systems with nanocrystals have an effect on the optical properties of the whole system. In this context, it is important to reveal the structural features of systems containing nanocrystals, characterizing aggregation, and adsorption processes in these systems, which makes it possible to establish a correlation between the structural and the optical properties of functional nanocomposite systems.

Silicon nanoclusters embedded in various transparent media are a new, interesting object for physicochemical investigation. For example, for particles smaller than 4 nm in size, quantum-size effects become significant. It makes possible to control the luminescence and absorption characteristics of materials based on such particles using of these effects [3, 4]. For nanoparticles about 10 nm in size or larger (containing $\sim 10^4$ Si atoms), the absorption characteristics in the UV and visible ranges are determined in many respects by properties typical of massive crystalline or amorphous silicon samples. These characteristics depend on a number of factors: the presence of structural defects and impurities, the phase state, etc. [5, 6]. For effective practical application and creation on a basis nc-Si, the new polymeric materials possessing useful properties: sun-protection films [7] and the coverings [8] photoluminescent and electroluminescent composites [9, 10], stable to light dyes [11], embedding of these nanosized particles in polymeric matrixes becomes an important synthetic problem.

The method of manufacture of silicon nanoparticles in the form of a powder by plasma chemical deposition, which was used in this study, makes possible to vary the chemical composition of their surface layers. As a result, another possibility of controlling their spectral characteristics arises, which is absent in conventional methods of manufacture of nanocrystalline silicon in solid matrices (e.g., in $\alpha-SiO_2$) by implantation of charged silicon particles [5] or radiofrequency deposition of silicon [2]. Polymer composites based on silicon nanopowder are a new object for comprehensive spectral investigation. At the same time, detailed spectral analysis has been performed for silicon nanopowder prepared by laser-induced decomposition of gaseous SiH_4 (see, for example, [6, 12]). It is of interest to consider the possibility of designing new effective UV protectors based on polymer-containing silicon nanoparticles [13]. An advantage of this nanocomposite in comparison with other known UV protectors is its environmental safety, that is, the ability to hinder the formation of biologically harmful compounds during UV-induced degradation of components of commercial materials. In addition, changing the size distribution of nanoparticles and their concentration in a polymer and correspondingly modifying the state of their surface, one can deliberately change the spectral characteristics of nanocomposite as a whole. In this case, it is necessary to minimize the transmission in the wavelength range below 400 nm (which determines the properties of UV-protectors [13]) by changing the characteristics of the silicon powder.

15.2 OBJECTS OF RESEARCH

In this study, the possibilities of using polymers-containing silicon nanoparticles as effective UV protectors are considered. First, the structure of nc-Si obtained under different conditions and its aggregates, their adsorption, and optical properties was studied to find ways of controlling the UV spectral characteristics of multiphase polymer composites containing nanocrystalline silicon. In addition, the purpose of this work was to investigate the effect of the concentration of silicon nanoparticles embedded in polymer matrix and the methods of preparation of these nanoparticles on the spectral characteristics of such nanocomposites. On the basis of the data obtained, recommendations for designing UV protectors based on these nanocomposites were formulated.

nc-Si consists of core-shell nanoparticles in which the core is crystalline silicon coated with a shell formed in the course of passivation of nc-Si with oxygen and/or nitrogen. nc-Si samples were synthesized by an original procedure in an argon plasma in a closed gas loop. To do this, we used a plasma vaporizer/condenser operating in a low-frequency arc discharge. A special consideration was given to the formation of a nanocrystalline core of specified size. The initial reagent was a silicon powder, which was fed into a reactor with a gas flow from a dosing pump. In the reactor, the powder vaporized at 7,000–10,000°C. At the outlet of the high-temperature plasma zone, the resulting gas–vapor mixture was sharply cooled by gas jets, which resulted in condensation of silicon vapor to form an aerosol. The synthesis of nc-Si in a low-frequency arc discharge was described in detail in [3].

The microstructure of nc-Si was studied by transmission electron microscopy (TEM) on a Philips NED microscope. X-ray powder diffraction analysis was carried out on a Shimadzu Lab XRD-6,000 diffractometer. The degree of crystallinity of nc-Si was calculated from the integrated intensity of the most characteristic peak at $2\theta = 28°$. Low-temperature adsorption isotherms at 77.3 K were measured with a Gravimat-4303 automated vacuum adsorption apparatus. FTIR spectra were recorded on in the region of 400—5,000 cm^{-1} with resolution of about 1 cm^{-1}.

Three samples of nc-Si powders with specific surfaces of 55, 60, and 110 m^2/g were studied. The D values for these samples calculated by Eq. (15.2) are 1.71, 1.85, and 1.95, respectively; that is, they are lower than the limiting values for rough objects. The corresponding D values calculated by Eq. (15.3) are 2.57, 2.62, and 2.65, respectively. Hence, the adsorption of nitrogen on nc-Si at 77.3 K is determined by capillary forces acting at the liquid—gas interface. Thus, in argon plasma with addition of oxygen or nitrogen, ultradisperse silicon particles are formed, which consist of a crystalline core coated with a silicon oxide or oxynitride shell. This shell prevents the degradation or uncontrollable transformation of the electronic properties of nc-Si upon its integration into polymer media. Solid structural elements (threads or nanowires) are structurally similar, which stimulates self-organization leading to fractal clusters. The surface fractal dimension of the clusters determined from the nitrogen adsorption isotherm at 77.3 K is a structurally sensitive parameter, which characterizes both the structure of clusters and the morphology of particles and aggregates of nanocrystalline silicon.

As the origin materials for preparation film nanocomposites served polyethylene of low density (LDPE) marks 10803-020 and ultradisperse crystal silicon. Silicon powders have been received by a method plazmochemical recondensation of coarse-crystalline silicon in nanocrystalline powder. Synthesis nc-Si was carried out in argon plasma in the closed gas cycle in the plasma evaporator the condenser working in the arc low-frequency category. After particle synthesis nc-Si were exposed microcapsulating at which on their surfaces the protective cover from SiO_2, protecting a powder from atmospheric influence and doing it steady was created at storage. In the given work, powders of silicon from two parties were used: nc-Si-36 with a specific surface of particles ~36 m^2/g and nc-Si-97 with a specific surface ~97 m^2/g.

Preliminary mixture of polyethylene with a powder nc-Si firms "Brabender" (Germany) carried out by means of closed hummer chambers at temperature 135 ± 5°C, within 10 min and speed of rotation of a rotor of 100 min^{-1}. Two compositions LDPE + nc-Si have been prepared: (1) composition PE + 0.5 per cent nc-Si-97 on a basis nc-Si-97, containing 0.5 weights silicon per cent; (2) composition PE + 1 per cent ncSi-36 on a basis ncSi-36, containing 1.0 weights silicon per cent.

Formation of films by thickness 85 ± 5 μ was spent on semi-industrial extrusion unit ARP-20-150 (Russia) for producing the sleeve film. The temperature was 120–190°C on zones extruder and extrusion die. The speed of auger was 120 min^{-1}. Technological parameters of the nanocomposites choose, proceeding from conditions of thermostability and the characteristic viscosity recommended for processing polymer melting.

15.3   EXPERIMENTAL METHODS

Mechanical properties and an optical transparency of polymer films, their phase structure, crystallinity, and also communication of mechanical and optical properties with a microstructure of polyethylene and granulometric structure of modifying powders nc-Si were observed.

Physicomechanical properties of films at a stretching (extrusion) measured in a direction by means of universal tensile machine EZ-40 (Germany) in accordance with Russian State Standard GOST-14236-71. Tests are spent on rectangular samples in width of 10 mm, and a working site of 50 mm. The speed of movement of a clip was 240 mm/min. The five parallel samples were tested.

Optical transparency of films was estimated on absorption spectra. Spectra of absorption of the obtained films were measured on spectrophotometer SF-104 (Russia) in a range of wavelengths 200–800 nm. Samples of films of polyethylene and composite films PE + 0.5 per cent nc-Si-36 and PE + 1 per cent nc-Si-36 in the size 3 × 3 cm were investigated. The special holder was used for maintenance uniform a film tension.

X-ray diffraction analysis by wide-angle scattering of monochromatic X-rays data was applied for research phase structure of materials, degree of crystallinity of a polymeric matrix, the size of single-crystal blocks in powders nc-Si and in a polymeric matrix, and also functions of density of distribution of the size crystalline particles in initial powders nc-Si.

X-ray diffraction measurements were observed on Guinier diffractometer: chamber G670 Huber [14] with bent Ge (111) monochromator of a primary beam, which are cutting out line $K\alpha_1$ (length of wave λ = 1.5405981 Å) characteristic radiation of X-ray tube with the copper anode. The diffraction picture in a range of corners 2θ from 3° to 100° was registered by the plate with optical memory (IP-detector) of the camera bent on a circle. Measurements were spent on original powders nc-Si-36 and nc-Si-97, on the pure film LDPE further marked as PE, and on composite films PE + 0.5 per cent nc-Si-97 and PE + 1.0 per cent nc-Si-36. For elimination of tool distortions effect diffractogram standard SRM660a NIST from the crystal powder LaB_6 certificated for these purposes by Institute of standards of the United States was measured. Further, it was used as diffractometer tool function.

Samples of initial powders nc-Si-36 and nc-Si-97 for X-ray diffraction measurements were prepared by drawing of a thin layer of a powder on a substrate from a special film in the thickness 6 μ (MYLAR, Chemplex Industries Inc., Cat. No: 250, Lot No: 011671). Film samples LDPE and its composites were established in the diffractometer holder without any substrate, but for minimization of structure effect two layers of a film focused by directions extrusion perpendicular each other were used.

Phase analysis and granulometric analysis was spent by interpretation of the X-ray diffraction data. For these purposes the two different full-crest analysis methods [15, 16] were applied: (1) method of approximation of a profile diffractogram using analytical functions, polynoms, and splines with diffractogram decomposition on making parts and (2) method of diffractogram modeling on the basis of physical principles of scattering of X-rays. The package of computer programs WinXPOW was applied to approximation and profile decomposition diffractogram ver. 2.02 (Stoe, Germany) [17], and diffractogram modeling at the analysis of distribution of particles in the sizes was spent by means of program PM2K (version 2009) [18].

15.4 RESULTS AND DISCUSSION

Results of mechanical tests of the prepared materials are presented in Table (15.1) from which it is visible that additives of particles nc-Si have improved mechanical characteristics of polyethylene.

TABLE 15.1 Mechanical characteristics of nanocomposite films based of LDPE and nc-Si

Sample	Tensile strength (kg/cm²)	Relative elongation-at-break (%)
PE	100 ± 12	200–450
PE + 1% nc-Si-36	122 ± 12	250–390
PE + 0.5% nc-Si-97	118 ± 12	380–500

The results presented in Table (15.1) shows that additives of powders of silicon raise mechanical characteristics of films, and the effect of improvement of mechanical properties is more expressed in case of composite PE + 0.5 per cent nc-Si-97 at which in comparison with pure polyethylene relative elongation-at-break has essentially grown.

Transmittance spectra of the investigated films are shown on Figure (15.1).

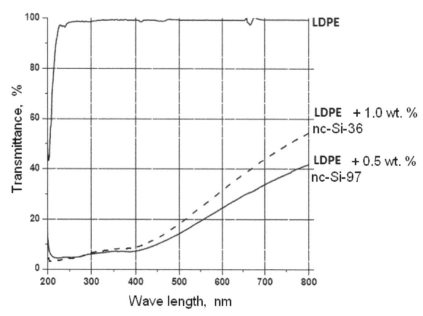

FIGURE 15.1 Transmittance spectra of the investigated films LDPE and nanocomposite films PE + 0.5 per cent nc-Si-97 and PE + 1.0 per cent nc-Si-36.

It is visible that additives of powders nc-Si reduce a transparency of films in all investigated range of wavelengths, but especially strong decrease transmittance (almost in 20 times) is observed in a range of lengths of waves of 220–400 nm, i.e., in UV areas.

The wide-angle scattering of X-rays data were used for the observing phase structure of materials and their component. Measured X-ray diffractograms of initial powders nc-Si-36 and nc-Si-97 on intensity and Bragg peaks position completely corresponded to a phase of pure crystal silicon (a cubic elementary cell of type of diamond—spatial group $Fd\bar{3}m$, cell parameter $a_{Si} = 0.5435$ nm).

For the present research, granulometric structure of initial powders nc-Si is of interest. Density function of particle size in a powder was restored on X-ray diffractogram a powder by means of computer program PM2K [18] in which the method [19] modeling of a full-profile diffractogram based on the theory of physical processes of diffraction of X-rays is realized. Modeling was spent in the assumption of the spherical form of crystalline particles and logarithmically normal distributions of their sizes. Deformation effects from flat and linear defects of a crystal lattice were considered. Received function of density of distribution of the size crystalline particles for initial powders nc-Si are represented graphically on Figure (15.2), in the signature to which statistical parameters of the found distributions are resulted. These distributions are characterized by such important parameters, as *Mo(d)*—position of maximum (a distribution mode); $<d>_V$—average size of crystalline particles based on volume of the sample (the average arithmetic size) and *Me(d)*—the median of distribution defining the size *d*, specifying that particles with diameters less than this size make half of volume of a powder.

The results represented in Figure (15.2) show that initial powders nc-Si in the structure have particles with the sizes less than 10 nm, which especially effectively absorb UV radiation. Both the powders modes of density function of particle size are very close, but median of density function of particle size of a powder nc-Si-36 is more essential than at a powder nc-Si-97. It suggests that the number of crystalline particles with diameters is less 10 nm in unit of volume of a powder nc-Si-36 much less, than in unit of volume of a powder nc-Si-97. As a part of a powder nc-Si-36 it is a lot of particles with a diameter more than 100 nm and even there are particles more largely 300 nm whereas the sizes of particles in a powder nc-Si-97 do not exceed 150 nm and the basic part of crystalline particles has diameter less than 100 nm.

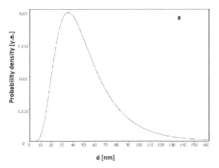 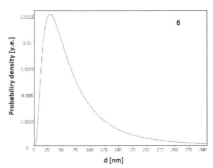

FIGURE 15.2 Density function of particle size in powders nc-Si, received from X-ray diffractogram by means of program PM2K: (a)—nc-Si-97 Mo(d) = 35 nm Me(d) = 45 nm $<d>_V$ = 51 nm; and (b)—nc-Si-36 Mo(d) = 30 nm Me(d) = 54 nm $<d>_V$ = 76 nm.

The phase structure of the obtained films was estimated on wide-angle scattering diffractogram only qualitatively. Complexity of diffraction pictures of scattering and structure does not pose the quantitative phase analysis of polymeric films [20]. At the phase analysis of polymers, often it is necessary to be content with the comparative qualitative analysis that allows watching evolution of structure depending on certain parameters of technology of production. Measured wide-angle X-rays scattering diffractograms of investigated films are shown on Figure (15.3). Diffractogramms have a typical form for polymers. As a rule, polymers are the two-phase systems consisting of an amorphous phase and areas with distant order, conditionally named crystals. Their diffractograms represent [20] superposition of intensity of scattering by the amorphous phase which is looking like wide halo on the small-angle area (in this case in area 2θ between 10° and 30°), and intensity Bragg peaks scattering by a crystal phase.

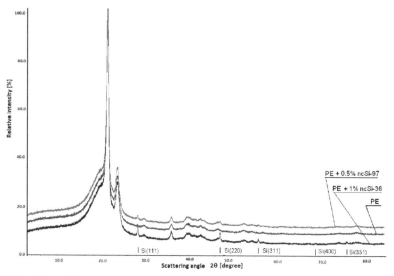

FIGURE 15.3 Diffractograms of the investigated composite films in comparison with diffractogram of pure polyethylene. Below vertical strokes specify reference positions of diffraction lines of silicon with their interference indexes (hkl).

Data on Figure (15.3) are presented in a scale of relative intensities (intensity of the highest peak is accepted equal 100%). For convenience of consideration, curves are represented with displacement on an axis of ordinates. The scattering plots without displacement represented completely overlapping of diffractogram profiles of composite films with diffractogram of a pure LDPE film, except peaks of crystal silicon, which were not present on PE diffractogram. It testifies that additives of powders nc-Si practically have not changed crystal structure of polymer.

The peaks of crystal silicon are well distinguishable on diffractograms of films with silicon (the reference positions with Miller's corresponding indexes are pointed below). Heights of the peaks of silicon with the same name (i.e., peaks with identical indexes) on diffractograms of the composite films PE + 0.5 per cent nc-Si-97 and PE + 1.0 per cent nc-Si-36 differ approximately twice that corresponds to a parity of mass concentration Si set at their manufacturing.

Degree of crystallinity of polymer films (a volume fraction of the crystal ordered areas in a material) in this research was defined by diffractograms (see Figure 15.3 for a series of samples only semiquantitative (more/less)). The essence of the method of crystallinity definition consists in analytical division of a diffractogram profile on the Bragg peaks from crystal areas and diffusion peak of an amorphous phase [20], as shown in Figure (15.4).

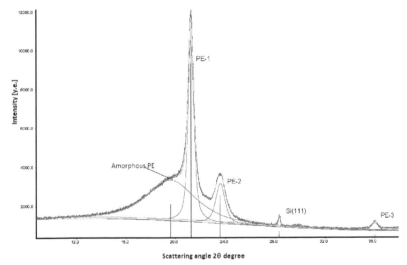

FIGURE 15.4 Diffractogram decomposition on separate peaks and a background by means of approximation of a full profile by analytical functions on an example of the data from sample PE + 1 per cent nc-Si-36 (Figure 15.3). PE-n designate Bragg peaks of crystal polyethylene with serial numbers n from left to right. Si (111)—Bragg silicon peak nc-Si-36. Vertical strokes specify positions of maxima of peaks.

Peaks profiles of including peak of an amorphous phase were ap-approximated by function pseudo-Voigt, a background four-order polynoms of Chebysheva. The nonlinear method of the least squares minimized a difference between intensity of points experimental and approximating curves. The width and height of approximating functions, positions of their maxima and the integrated areas, and also background parameters were thus specified. The relation of integrated intensity of a scattering profile by an amorphous phase to full integrated intensity of scattering by all phases except for particles of crystal silicon gives a share of amorphy of the sample, and crystallinity degree turns out as a difference between unit and an amorphy fraction.

It was supposed that one technology of film obtaining allowed an identical structure. It proved to be true by coincidence relative intensities of all peaks on diffractograms Figure (15.3), and samples consist only crystal and amorphous phases of the same chemical compound. Therefore, received values of degree of crystallinity should reflect correctly a tendency of its change at modification polyethylene by powders nc-Si though because of a structure of films they can quantitatively differ considerably from the valid concentration of crystal areas in the given material. The found values of degree of crystallinity are represented in Table (15.2).

TABLE 15.2 Characteristics of the ordered (crystal) areas in polyethylene and its composites with nc-Si

PE			PE + 1% nc-Si-36			PE + 0.5% nc-Si-97		
Crystallinity	**46%**		**47.5%**			**48%**		
2θ [°]	d [E]	ε	2θ [°]	d [E]	ε	2θ [°]	d [E]	e
21.274	276	8.9	21.285	229	7.7	21.282	220	7.9
23.566	151	12.8	23.582	128	11.2	23.567	123	11.6
36.038	191	6.8	36.035	165	5.8	36.038	162	5.8
Average values	206	9.5×10^{-3}		174	8.2×10^{-3}		168	8.4×10^{-3}

One more important characteristic of crystallinity of polymer is the size d of the ordered areas in it. For definition of the size of crystalline particles and their maximum deformation ε *in* X-ray diffraction analysis [21], Bragg peaks width on half of maximum intensity (Bragg lines half-width) is often used. In the given research, the sizes of crystalline particles in a polyethylene matrix calculated on three well expressed diffractogram peaks (Figure 15.3). The peaks of polyethylene located at corners 2θ approximately equal 21.28°, 23.57°, and 36.03° (see peaks PE-1, PE-2 and PE-3 on Figure 15.4) were used. The ordered areas size d and the maximum relative deformation ε of their lattice were calculated by the joint decision of the equations of Sherrera and Wilson [21] with the use of half-width of the peaks defined as a result of approximation by analytical functions, and taking into account experimentally measured diffractometer tool function. Calculations were spent by means of program WinX^{POW} size/strain. Received d and ε, and also their average values for investigated films are presented in Table (15.2). The updated positions of maxima of diffraction peaks used at calculations are specified in the table.

The technology offered allowed the obtaining of films LDPE and composite films LDPE + 1 per cent nc-Si-36 and LDPE + 0.5 per cent nc-Si-97 an identical thickness (85 μ). Thus concentration of modifying additives nc-Si in composite films corresponded to the set structure that is confirmed by the X-ray phase analysis.

By direct measurements, it is established that additives of powders nc-Si have reduced a polyethylene transparency in all investigated range of lengths of waves, but especially strong transmittance decrease (almost in 20 times) is observed in a range of lengths of waves of 220–400 nm (i.e. in UV areas). Especially strongly, effect of suppression UV of radiation is expressed in LDPE film + 0.5 per cent nc-Si-97 although concentration of an additive of silicon in this material is less. It is possible to explain this fact to that according to experimentally received function of density of distribution of the size the quantity of particles with the sizes is less 10 nm on volume/weight unit in a powder nc-Si-97 more than in a powder nc-Si-36.

Direct measurements define mechanical characteristics of the received films—durability at a stretching and relative lengthening at disrupture (Table 15.1). The received results show that additives of powders of silicon raise durability of films approximately on 20 per cent in comparison with pure polyethylene. Composite films in comparison with pure polyethylene also have higher lengthening at disrupture, especially this improvement is expressed in case of composite PE + 0.5 per cent nc-Si-97. Observable improvement of mechanical properties correlates with degree of crystallinity of films and the average sizes of crystal blocks in them (Table 15.2).

By results of the X-ray analysis the highest crystallinity at LDPE film + 0.5 per cent nc-Si-97, and at it the smallest size the crystal ordered areas that should promote durability and plasticity increase.

This work is supported by grants RFBR no 10-02-92000 and RFBR no 11-02-00868 also by grants FCP "Scientific and scientific and pedagogical shots of innovative Russia," contract no 2353 from November 17, 2009 and contract no 2352 from November 13, 2009

KEYWORDS

- **Nanocrystalline silicon**
- **Polyethylene**
- **Polymer nanocomposites**
- **Spectroscopy**
- **UV-protective film**
- **X-ray diffraction analysis**

REFERENCE

1. Karpov, S. V.; and Slabko, V. V.; Optical and Photophysical Properties of Fractally Structured Metal Sols. Novosibirsk: Sib. Otd. Ross. Akad. Nauk; **2003**.
2. Varfolomeev, A. E.; Volkov, A. V.; and Godovskii, D. Yu. et al. *Pis 'ma Zh. Eksp. Teor. Fiz.* **1995**, *62*, 344.
3. Delerue, C.; Allan, G.; and Lannoo, M.; *J. Lumin.* **1999**, *80*, 65.
4. Soni, R. K.; Fonseca, L. F.; and Resto, O. et al. *J. Lumin.* **1999**, *187*, 83–84.
5. Altman, I. S.; Lee, D.; and Chung, J. D. et al. *Phys. Rev. B: Condens. Matter Mater. Phys.* **2001**, *63*, 161402.
6. Knief, S. and von Niessen, W.; *Phys. Rev. B: Condens. Matter Mater. Phys.* **1999**, *59*, 12940.
7. Olkhov, A. A.; Goldschtrakh, M. A.; and Ischenko, A. A.; RU Patent No 2009145013. **2009**.
8. Bagratashvili, V. N.; Tutorskii, I. A.; and Belogorokhov, A. I.; et al. Reports of Academy of Sciences. Physical Chemistry; **2005**, *405*, 360 p.
9. Kumar, V. Ed. Nanosilicon. Elsevier Ltd; **2008**, *xiii+*, 368 p.
10. Olkhov, A.; Nanostructured materials. In: Processing, Properties, and Applications. Ed. Koch, Carl C.; New York: William Andrew Publishing; **2009**, 752 p.
11. Ischenko, A. A.; Dorofeev, S. G.; and Kononov, N. N.; et al. RU Patent No 2009146715. **2009**.

12. Kuzmin, G. P.; Karasev, M. E.; and Khokhlov, E. M.; et al. *Laser Phys.* **2000,** *10,* 939.

13. Beckman, J.; and Ischenko, A. A.; RU Patent No 2 227 015. **2003.**

14. Stehl, K.; The huber G670 imaging-plate Guinier camera tested on beamline I711 at the MAX II synchrotron. *J. Appl. Cryst.* **2000,** *33,* 394–396.

15. Fetisov, G. V.; The X-ray Phase Analysis. Chapter 11, 153–184 p. Analytical Chemistry and Physical and Chemical Methods of the Analysis. T. 2. Red. Ischenko, A. A.; M.: ITc Academy; **2010,** 416 p.

16. Scardi, P.; and Leoni, M.; Line profile analysis: pattern modelling versus profile fitting. *J. Appl. Cryst.* **2006,** *39,* 24–31.

17. Olkhov, A.; WINX^{POW} Version 1.06. STOE & CIE GmbH Darmstadt/Germany; **1999.**

18. Leoni, M.; Confente, T.; and Scardi, P.; PM2K: a flexible program implementing whole powder pattern modelling. *Z. Kristallogr. Suppl.* **2006,** *23,* 249–254.

19. Scardi, P.; Recent advancements in whole powder pattern modeling. *Z. Kristallogr. Suppl.* **2008,** *27,* 101–111.

20. Strbeck, N.; X-ray Scattering of Soft Matter. Berlin Heidelberg: Springer-Verlag; **2007,** *xx+,* 238 p.

21. Iveronova, V. I.; and Revkevich, U. P.; The Theory of Scattering of X-rays. M.: MGU; **1978,** 278 p.

CHAPTER 16

A CASE STUDY ON BIODEGRADABLE COMPOSITIONS BASED ON NATURAL POLYMERS

S. G. KARPOVA, A. L. IORDANSKII, A. A. POPOV,
S. M. LOMAKIN, and N. G. SHILKINA

CONTENTS

16.1   INTRODUCTION

Biodegradable compositions based on natural polymers are widely used in practice. Because of mixing, new physicochemical properties of compositions can arise, which are not inherent to the original components. Innovative technologies use biodegradable system for producing clean construction materials and packagings [4, 5, 10, 11, 12,]. A typical representative of polyhydroxyalkanoates [6] is polyhydroxybutyrate (PHB), along with its useful properties has some undesirable characteristics: high cost and fragility. To overcome these limitations its copolymers with poly-(3 hydroxyvalerate) (PHBV), as well as compositions with other biomedical polymers, in particular chitosan. Chitosan, a natural non-toxic biodegradable polymer, produced by deacetylation of chitin, is the second most common biopolymer in nature after cellulose. Chitosan is used in film coatings and membranes to extend the shelf life of foodstuffs. However, a high sensitivity chitosan to moisture limits its applications. This disadvantage can be overcome by mixing chitosan with moisture proof polymers. In this case, however, the material remains biodegradable.

Varying the composition of the PHB/chitosan mixture and thus affecting its morphology and crystallinity makes it possible to prepare composite materials with various physicochemical characteristics, such as permeability, water solubility, the rate, and mechanism of degradation, etc. In [8], differential scanning calorimetry (DSC) and NMR spectroscopy measurements showed that the crystallization of PHB in blends with chitosan is increasingly suppressed with increasing concentration of polysaccharides. The authors supposed that can form hydrogen bonds between carbonyl groups of PHB and amide groups of chitosan. However, there is evidence [9] on the changes of PHB in PHB/chitosan blends that contradicts the results of [8], which is likely associated with mixture preparation conditions.

An effective way to assess the state of the amorphous and crystalline phases of original polymers and mixtures thereof is to use a combination of dynamic and structural methods. In this study, we used EPR spectroscopy (probe analysis) DSC, and large angle XRD, and UV spectroscopy. This combination of structural and dynamic methods enable to obtain a more complete assessment of the structural evolution of the PHB/chitosan mixture in an aqueous medium over short time intervals (a few hours), which precedes the hydrolytic decomposition of the polymer system.

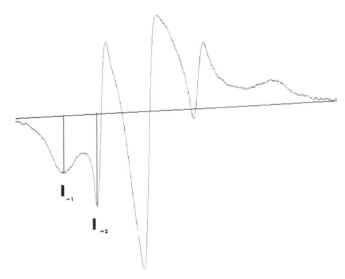

FIGURE 16.1 EPR spectrum of a mixed composition containing 80 per cent chitosan and 20 per cent PHB.

16.2 OBJECTS AND METHODS

We studied mixtures of biodegradable polymers, polyhydroxybutyrate and chitosan. Films were prepared from natural biodegradable polymer polyhydroxybutyrate of series 16F, produced by microbiological synthesis (BIOMER®, Germany). Original polymer was a fine white powder. Poly-hydroxybutyrate has a molar mass of $Mw = 2.06 \cdot 10^5$ g/mol, density of $d = 1.248$ g/cm3, $Tm = 177°C$, and degree of crystallinity of 65 per cent. Chitosan, of domestic production (Bioprogress, Shchelkovo), an infusible polysaccharide, was in the form of a fine powder. The molar weight of this polymer is $Mw = 4.4 \cdot 10^5$ g/mol and the degree of deacetylation, 82.3 per cent. In casting the films, we used the following solvents: chloroform and dioxane (reagent grade) (ZAO Ekos, Russia) for PHB and acetic acid (reagent grade) for chitosan. The films were prepared by mixing a solution of chitosan in an aqueous acidic medium and a PHB solution in dioxane. The chitosan solution was prepared by dissolving its powder in acetic acid.

Molecular mobility was studied by the paramagnetic probe method, by measuring the correlation time τ that characterizes the rotational mobility of the probe as described in Ref. [3]. The probe was the 2,2,6,6 tetramethylpiperidine 1 oxyl (TEMPO) radical, introduced into the film in the form of its vapors.

Calorimetric studies of the samples were performed on a DSK 204 F1 instrument (Netzsch, Germany) in an inert atmosphere of argon at a heating rate of 10 K/min. The average measurement error was 2 per cent.

X ray diffraction analysis of the samples was carried out on RU 200 Rotaflex instrument (Rigaku, Japan) with a 12 kW generator with a rotating copperanode in the transmission mode (40 kV, 140 mA, Cu($K\alpha$) radiation ($\lambda = 0.1542$ nm, Ni filter).

The kinetics of the release of rifampicin was studied using a DU 65 UV spectrophotometer (Backman USA) with output to a chart recorder. For this purpose, a rifampicin film was immersed in a container with a phosphate buffer solution (pH 6.86), and the kinetics of rifampicin release was monitored by periodic sampling. The measurement error was 5 per cent.

16.3   RESULTS AND DISCUSSION

16.3.1   *STUDYING THE INTERACTION OF THE AMORPHOUS PHASE OF PHB AND ITS COMPOSITIONS WITH CHITOSAN WITH WATER*

EPR spectra of chitosan-PHB mixed compositions and of the individual component are superpositions of two spectra corresponding to radicals with correlation times t_1 and t_2: t_2, which characterize molecular mobility in loose amorphous regions and in a more dense amorphous phase, respectively (Figure 16.1). All calculations of EPR spectra were performed using the stochastic Liouville equation (random trajectory method) [1]. Let and be the intensities of the first peaks of the triplets corresponding hindered and faster motion, respectively. We plotted the dependence of the ratio of the intensities of the first lines of the spectra, (I_{+1}/I_{+}) on the composition of the mixture (Figure 16.2). This parameter characterizes the ratio of the fraction of amorphous phase with slow molecular mobility to that of amorphous phase with fast mobility. It is seen that, in pure PHB, the fraction of slow component is small, increasing by almost an order of magnitude in mixed compositions. It should be noted that, in pure chitosan and mixed compositions with 10 and 20 per cent PHB, this ratio is even higher. These data are indicative of a heterogeneous structure, which is especially pronounced in mixed compositions and chitosan.

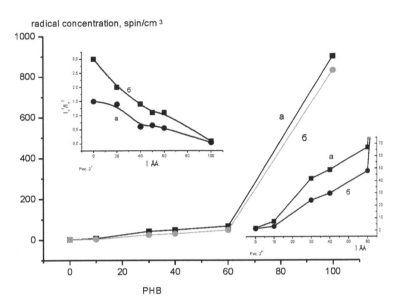

FIGURE 16.2 Dependence of the concentration of TEMPO radicals in samples of the same mass on the composition of the mixture; (a) ratio of the first peaks of the spectra with correlation times t_1 and t_2 (1) before and (2) annealing. (b) Concentration of the radical in mixed compositions of the same mass with PHB content of up to 60 per cent. An experiment was performed to determine the number of radicals in identical samples of PHB, chitosan, and mixtures thereof. Figure (16.2) shows that the number of radicals in pure PHB is almost two orders of magnitude larger than that in chitosan and compositions. These results indicate a low permeability of interphase regions for the radical. Note also that, starting from 30 per cent PHB, the number of radicals in samples (Figure 16.2(b)), is almost 10 times higher than in mixtures with 10 and 20 per cent PHB, which means that chitosan and mixtures containing up to 20 per cent PHB (inclusive) have the densest structure of amorphous regions. The shape of the spectra of mixed compositions and chitosan is very different from the spectrum of PHB. While the first spectrum of PHB, representing hindered motion, is almost superimposed on the spectrum corresponding to fast motion, the respective spectra of chitosan and mixed compositions are clearly separated (Figure 16.1). These data suggest that the continuous phase in mixed compositions is chitosan and that the interphase layers have a relatively high density. Figure (16.3) (curve 1) shows how the correlation time t_2 depends on the composition of the mixture. As can be seen, in individual chitosan and in compositions with 10 and 20 per cent PHB, the correlation time is relatively short (9.5ч19) 10^{-10} s, while compositions with 30 per cent PHB are characterized by a correlation time of 280 10^{-10} s (for PHB, $t_2 = 65\ 10^{-10}$ s). Other compositions also have long correlation times. According to DSC data, the glass transition temperature of chitosan is ≈40°C, whereas a mixture with 30 per cent PHB features no glass transition peak in the entire temperature range covered (from −20 to 200°C). The results suggest that, when mixed with PHB, chitosan experiences a transition from the glassy to a highly elastic state, so that the radical, penetrating into dense regions of the phase, provides high values of the correlation time t.

The samples were exposed to water at 70°C for 1 h. Curve 2 in Figure (16.3) shows how the correlation time depends on the composition of the mixture. It can be seen that the behavior of the dependence is the same only τ_2 increases, indicating a decline in molecular mobility in the loose amorphous phase. The ratio of the intensities of the first lines in the spectrum (/) with correlation times τ_1 and τ_2 after annealing in water (Figure 16.2(a), curve 2) indicates an increase in the fraction of amorphous regions with low mobility. Only PHB showed no difference. Note also that the number of radicals sorbed in compositions after exposure to water at 70°C for 1 h was smaller than in the initial polymers, which also indicates an increase in chain rigidity upon thermal treatment in water (Figure 16.2).

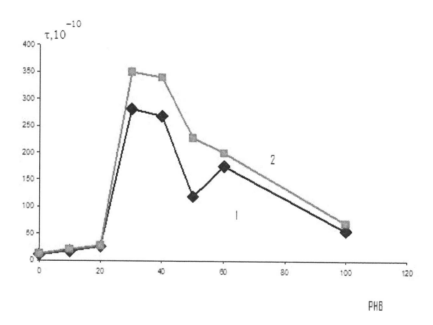

FIGURE 16.3 Dependence of the correlation time on the composition of the PHB-chitosan mixture (1) before and (2) after annealing at 70°C for 1 h.

The values of the activation energy for all the studied compositions are given in the table. Figure (16.4) shows the temperature dependence of the correlation time for three compositions. As can be seen, for PHB, the dependence is linear, whereas those for the mixed compositions exhibit a kink at ≈60°C. This means that at this temperature, molecular mobility in dense regions of chitosan is "unfrozen," since the activation energy E is almost 2 times higher than that for PHB.

TABLE 16.1 Activation energies for mixed compositions of chitosan and PHB

Content of chitosan (%)	E_1 kJ/mol	E_2 kJ/mol
100	7.5	
90	8.8	42
80	10	38
70	33	50
60	31	45
50	9	50
40	25	60
0	27	27

Note: The experimental error is 5 per cent.

Note that, in the initial segment, the value of E for chitosan and mixed compositions with 10 and 20 per cent of PHB is small, increasing almost 2.5-fold for compositions with a high content of PHB (except for a composition with 50% PHB). This also shows that, starting from 30 per cent PHB, the structure of the amorphous phase of mixed compositions becomes more rigid due to the transformation of the crystal structure into an amorphous (as shown below), and the radical, accumulating in these dense regions, indicates a slowdown of molecular motion.

We performed experiments aimed at studying the kinetics of the release of rifampicin from films. These studies are of great interest for solving problem concerning the diffusion of drugs. The rate of these processes largely depends on the structure of the amorphous phase of the polymer. The diffusion coefficients of rifampicin in mixtures was calculated by the equation $D = \pi l^2 (tg\alpha)^2 / 16$, where D is the diffusion coefficient, α is the angle of slope of the kinetic curves for the release of rifampicin from the film, and l is the film thickness. Figure (16.5) shows the dependence of D on the content of PHB in the composition. It is seen that, with increasing content of PHB in the composition of the system, the diffusion mobility declines, which may be due to a slowdown of the molecular mobility, as shown in Figure (16.2). Note that, just after the PHB content.

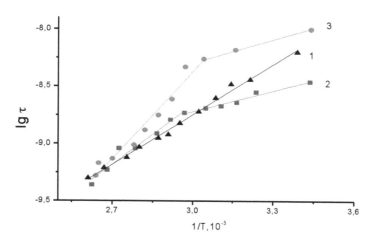

FIGURE 16.4   Temperature dependence of the correlation time for (1) PHB and mixed compositions with, (2) 20, and (3) 50 per cent PHB.

In the mixture becomes larger than 20 per cent, the diffusion coefficient decreases sharply, whereas for compositions with 30 and 40 per cent PHB, these values are close to each other (Figure 16.5). Thus, studying the amorphous phase in PHB-chitosan compositions showed that a mixture containing more than 20 per cent of PHB is characterized by the absence of the glassy component and has a relatively high density. The diffusion coefficient decreases with increasing PHB content in the composition, especially sharply in passing from 20 to 30 per cent PHB. Exposure to water at 70°C leads to an increase in molecular mobility due to compaction of the amorphous phase of the composition.

16.3.2   A STUDY OF THE CRYSTALLINE PHASE OF PHB, CHITOSAN, AND MIXTURES THEREOF

The crystalline phase of mixed compositions was studied using DSC (Figure 16.6). Characteristically, the bimodal shape of the peak of PHB melting transforms into a single peak for mixed compositions, which indirectly indicates the merger of the two forms of the crystalline state of PHB. The high-temperature peak at 175°C manifests the melting of the well orga-nized crystal structure, whereas the low-temperature peak at 161°C corre-sponds to the melting of the less ordered crystalline structure of PHB. DSC data showed that, with increasing PHB content in the mixture, its crystal-linity, and the melting temperature of the crystallites decrease. Thus, in compositions with 10 per cent PHB, the degree of crystallinity is as high as 82 per cent (Figure 16.6), but then, with increasing the PHB content, it decreased to 60 per cent (for PHB, $\chi = 61\%$). Note that the degree of crys-tallinity also decreases abruptly. While the crystallinity of a mixture with 20 per cent PHB is 79 per cent, in a 30 per cent PHB mixture, drops to $\chi = 60$ per cent, decreasing further $\chi = 56.7$ at 60 per cent PHB. According to DSC data, the degree of crystallinity of chitosan is low, only 6.5 per cent, with no crystalline phase detected in mixtures. The melting point of PHB in a mixed composition with 10 per cent PHB is 172.7°C, lowing to 171°C at 30 and 60 per cent PHB.

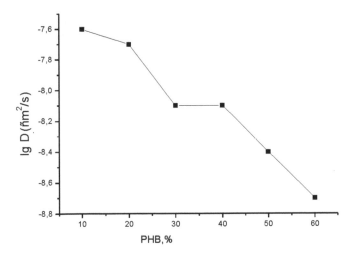

FIGURE 16.5 Dependence of the diffusion coefficient D of rifampicin on the composition of the PHB-chitosan mixture.

These data indicate that, when mixed with chitosan, PHB acquires a less ordered crystalline structure. This can be attributed to the interaction between chitosan and PHB via hydrogen bonds between ester groups of

PHB and amino groups of the polysaccharide. As a result, the structure of chitosan becomes looser, and the crystalline phase is non-existent in mixed compositions. It is important to note that, in mixed compositions with 10 and 20 per cent PHB, the crystallinity of the latter is extremely high, probably because of a low degree of entanglement of chains, a factor that facilitates crystallization, while the fraction hydrogen bonds remains very small. Similar results were obtained by using XRD. This method showed that the degree of crystallinity in chitosan is 38 per cent and decreases with increasing PHB content. For example, in a mixed composition with 20 per cent PHB, it is 30 per cent, decreasing to 15 per cent at 30 per cent PHB and to zero at 40 per cent PHB (Figure 16.7).

The crystallinity of PHB also decreases as its content in the mixed composition increase. The crystallinity of pure PHB is 62 per cent, whereas in mixtures chitosan, its crystallinity decreases to 54, 44, and 40 per cent at 20, 30, 40 per cent PHB, respectively. These data also show that the crystallinity of PHB in mixtures decreases sharply staring from its content of 30 per cent. Thus, with increasing PHB fraction (up to 60%) in mixtures with chitosan, the degree of crystallinity of both chitosan and PHB decline, indicating an increase in the fraction of dense regions in the amorphous phase. Note that, while the crystallinity of chitosan decreased by 5 per cent relative to pure chitosan upon mixing with 20 per cent PHB, in a mixture with 30 per cent PHB, the decrease reaches 23 per cent, with these changes being reflected in the correlation time. Note that, starting namely from 30 per cent PHB in the composition, the correlation time begins to increase sharply. This suggests that a fraction of the crystallites transforms into an amorphous state, and probe radicals in these mixtures, accumulating in these dense formations, provide a high correlation time.

According to DSC, exposure to water at 70°C for 1 h results in an increase in the degree of crystallinity of mixed compositions with 10, 20, 50, and 60 per cent PHB. In samples with 30 and 40 per cent PHB and in chitosan, the crystallinity decreased by almost half (Figure 16.6). According to XRD analysis, after exposure of mixed compositions with 10, 20, 30, and 40 per cent PHB to water at 70°C for 1 h, no crystalline phase in chitosan was detected. After such treatment, the melting point of PHB increased by 6–7°C for all the studied polymers. That the crystallinity of polymers increases after annealing is well known [2, 13], especially in view of the plasticizing effect of water, which accelerates crystallization. At the same time, the decrease of crystallinity with increasing PHB in the mixture is indicative of the interaction of chitosan with PHB via the formation of hydrogen bonds between ester group of PHB and amino groups of chitosan. Note that, after exposure to water at 70°C, chitosan in mixed compositions shows no crystallinity, as evidenced by XRD data.

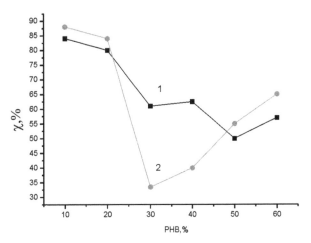

FIGURE 16.6 Dependence of the degree of crystallinity on the composition of the PHB-chitosan mixture (1) before and (2) after annealing at 70°C for 1 h.

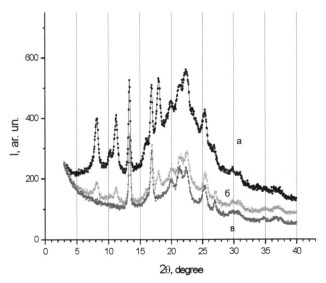

FIGURE 16.7 X-Ray diffract grams for PHB-chitosan mixtures with (1) 20, (2) 30, and (3) 40 per cent PHB.

16.4 CONCLUSIONS

Thus, it was shown that the structure of mixed compositions changes drastically stating from a 30 per cent PHB content (70% chitosan), which results in a significant increase in the correlation time. This is indicative of an increase in the fraction of structures with tightly packed chains, as confirmed by data on the diffusion of drugs. The degree of crystallinity also reduces sharply with increasing content of PHB in the composition, as demonstrated by DSC and XRD data. Thermal treatment at 70°C for 1 h causes an increase in the correlation time, indicative of a decrease in molecular mobility. At the same time, the degree of crystallinity changes in a complex manner.

KEYWORDS

- **Biodegradable composition**
- **Crystallinity**
- **DSC**
- **EPR probe**
- **Ozone**
- **Water**
- **XRD**

REFERENCE

1. Antsiferova, L. I.; Vasserman, A. M.; Ivanova, A. I.; Livshits, V. A.; and Nazemets, N. S.; Spectra Atlas. Moscow: Nauka; **1977**, [in Russian].
2. Artsis, M. I.; Bonartsev, A. P.; Iordanskii, A. L.; Bonartseva, G. A.; and Zaikov, G. E.; "Biodegradation and medical application of microbial poly(3-hydroxybutyrate)." *Mol. Cryst. Liq. Cryst.* **2010,** *523,* 21.
3. Buchachenko, A. L.; and Vasserman, A. M.; Stable Radicals. Moscow: Khimiya, **1973,** [in Russian].
4. Iordanskii, A. L.; Rogovina, S. Z.; Kosenko, R. Yu.; Ivantsova, E. L.; and Prut, E. V.; "Development of f biodegradable polyhydroxybutyrate-chitosan-rifampicin composition for controlled transport of biologically active compounds." *Dokl. Phys. Chem.* **2010,** *431,* 60.

5. Ivantsova, E. L.; Iordanskii, A. L.; Kosenko, R. Yu.; et al. "Structure and prolonged transport in a bio-degradable poly(R-3-hydroxybutyrate)-drug system." *Khim. Farm. Zh.* **2011,** *45,* 39.

6. Ivantsova, E. L.; Kosenko, R. Yu.; Iordanskii, A. L.; et al. "Structure and prolonged transport in a bio-degradable poly(R-3-hydroxybutyrate)-drug system." *Polym. Sci. Ser.* A **2012,** *54,* 87.

7. Nikolenko, Yu. M.; Kuryavyi, V. G.; Sheveleva, I. V.; Zemskova, L. A.; and Sergienko, V. I.; "The study by atomic force microscopy and X-ray photoelectron spectroscopy fibrous chitosan-carbon materials." *Inorg. Mater.* **2010,** *46,* 221.

8. Ikejima, T.; and Inoue, Y.; "Crystallization behavior and environ mental biodegradability of the blend films of poly(3-hydroxybutyric acid) with chitin and chitosan." *Carbohydr. Polym.* **2000,** *41,* 351.

9. Ikejima, T.; Yagi, K.; and Inoue, Y.; "Thermal properties and crystallization behavior of poly(3-hydroxybutyric acid) in blends with chitin and chitosan." *Macromol. Chem. Phys.* **1999,** *200,* 413.

10. Freier, T.; Sternberg, S.; Behrend, D.; and Schmitz, K. P.; In: Health Issues of Biopolymers. Polyhydroxybutyrate Series of Biopolymers. Ed. Steinbchel, A. V.; Wiley: Weinheim; **2003,** *10,* 247.

11. Polymer Blends, Handbook. Ed. Utracki, L. A.; Kluwer Academic, Dodrecht, **2002,** 1.

12. Shibryaeva, L. S.; Shatalova, O. V.; Krivandin, A. V.; et al. "Thermal oxidation of isotactic polypropy-lene synthesized with a metallocene catalyst." *Polym. Sci. Ser.* A **2003,** *45,* 244.

13. Yu, L.; Dean, K.; and Li, L.; "Polymer blends and composites from renewable resources." *Prog. Polym. Sci.* **2006,** *31,* 576.

CHAPTER 17

A RESEARCH NOTE ON THE INFLUENCE OF HYDROPHILIC FILLER CONTENT ON THE FIRE RESISTANCE OF COMPOSITES BASED ON EPOXY RESIN ED-20

V. F. KABLOV, A. A. ZHIVAEV, N. A. KEIBAL, T. V. KREKALEVA, and A. G. STEPANOVA

CONTENTS

17.1   INTRODUCTION

Today, expanding the range of technological and operational characteristics of composite materials based on epoxy oligomers is an urgent problem. Polymer composites based on epoxy resins are widely used as structural materials and adhesives. The advantages of epoxy composites are good adhesion to reinforcing elements, the lack of volatile by-products during hardening, and low shrinkage [1–2].

However, in some cases, the use of epoxy composites is limited by their low thermal stability and fire resistance [3]. One of the benefits of epoxy resins is ability to regulate their composition by introducing various modifiers (fillers, plasticizers, fire retardants, etc.) resulting in materials with a given set of properties [4]. There are known fire retardant polymer compositions with microencapsulated fire extinguishing liquids (halogen phosphorus containing compounds, water, etc.). Microencapsulation substantially improves both technological and functional properties of the most diverse products and considerably expands the scope of their application.

17.2   EXPERIMENTAL PART

In the research, the influence of hydrophilic filler content on the fire resistance of composites based on epoxy resin ED-20 was investigated.

Acrylamide copolymer POLYSWELL was applied as a hydrophilic filler. The acrylamide copolymer is a white granular material, density is 0.8–1.0 g/cm^3, water-swellable to form a polymer gel. In solutions the amide group shows weak-basic properties at the expense of the lone electron pair on the nitrogen atom that is the reason of non-chemical interaction of the polymer with water.

The compositions were obtained on the basis of epoxy resin by means of blending components as follows: epoxy resin ED-20, filler—acrylamide copolymer in the form of granules preliminary swollen in water in the ratio of 1:10, and hardener that was polyethylenepolyamine. The obtained reactive blends were molded and then hardened without heat supply for 24 h. The test samples had the following sizes: diameter is 50 mm, thickness is 5 mm.

For determining the efficiency of the developed composites, the experiments on the fire resistance by exposure of a sample to open flame using the universal Bunsen burner were conducted. With a help of the pyrometer C-300.3, measuring the moment of achieving the limit state, time-temperature transformations on the non-heated surface of the sample were registered.

Along with the fire resistance estimate of the developed compositions, the studies of samples on the water absorption and combustibility depending on the hydrophilic filler content were carried out.

The combustibility was evaluated by the standard technique on the rate of horizontal flame spread over the surface. A sample was exposed to the burner flame (temperature peak 840°C) and the burning and smoldering time after fire source elimination was fixed.

The experiments on the water absorption were performed in distilled water at temperature $23 \pm 2°C$ for 24 h. The water absorption was characterized by a sample weight change before and after exposure to water.

17.3   RESULTS AND DISCUSSION

As was mentioned above, the investigation on determining the fire resistant properties of epoxy compositions was carried out during the research. The results are shown in Figures (17.1) and (17.2).

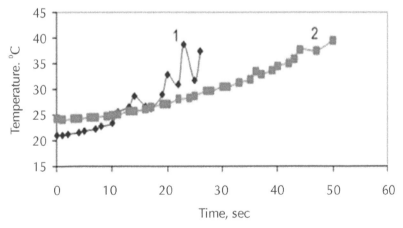

FIGURE 17.1   Dependence of temperature on the non-heated sample side on flame exposure time: (1)—initial epoxy component; and (2)—epoxy composite containing hydrophilic filler.

As it is seen in Figure (17.1), the test sample damage takes place on the 15th second, which is evidenced by temperature fluctuations on curve 1. The filled sample (contains 15% of the hydrophilic filler) maintains the integrity up to 50 sec; the sparking is observed at combustion that is, probably, related to water injection into the combustion zone. Besides, when the flame source is eliminated, the sample self-extinguishes for 2–3 sec.

The effect of hydrophilic filler content (5–20%) on the fire resistance of the composites was also studied in the work (Figure 17.2).

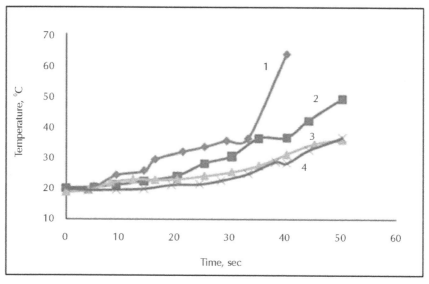

FIGURE 17.2 Dependence of temperature on the non-heated sample side on flame exposure time for epoxy composite containing hydrophilic filler in the following amounts: 5 per cent (1), 10 per cent (2), 15 per cent (3), and 20 per cent (4).

When measuring temperature on the nonheated surface of water-containing composites within a specified time span, it was established that the fire resistant properties improve with increasing hydrophilic filler content from 10 to 20 per cent; the sample with filler content of 5 per cent was damaged by 33 sec.

The results of experiments on the water absorption and fire resistance are presented in Table (17.1).

TABLE 17.1   Estimation of water absorption and fire resistance of epoxy composites

Parameter	Filler content in a composite, % by epoxy resin weight				
	Without filler	5	10	15	20
	Composition index				
Water absorption, %	0.41	0.33	0.24	0.28	0.28
Time to the limit state, sec	15.0	35.0	50.0	50.0	50.0
Temperature of the non-heated sample side in 25 sec, °C	Sample damaged	33.0	28.0	24.0	20.0

Proposed compositions provide a significant increase in values of the fire resistance and water absorption compared to the initial sample. Time to the limit state for the test samples goes up by 2.5 times at that.

The data obtained in combustibility tests of water-containing compositions are illustrated in Figure (17.3). The best results were obtained with filler content of 15 and 20 per cent. In this way, the flame spread rate for the initial sample is 18 mm/min and, when the hydrophilic filler is used, it is equal to 3 mm/min.

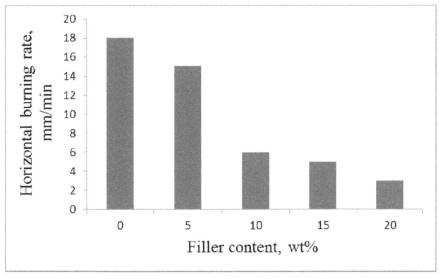

FIGURE 17.3   Evaluation of the horizontal flame spread rate over epoxy composite surface.

On exposure to flame, there is a kind of microexplosions and injection of fire extinguishing liquid–water occurring in the combustion zone. In this case, combustion inhibition is likely due to the absorption of a significant heat amount characterized by the high heat capacity and high water evaporation heat. A possible factor in reducing the flame spread rate is also water displacement of the combustion reaction components from the reaction zone.

17.4　CONCLUSION

So, the properties of water-containing epoxy composites were developed and investigated as well as the possibility of applying a hydrophilic filler as an additive that increase the fire resistance of hardened epoxy compositions based on epoxy resin ED-20 was shown in the paper.

One of the important conclusions in this work is the high prospect of the modification method of epoxy composites in order to give them a set of specific properties using filled microencapsulated materials.

KEYWORDS

- **Acrylamide copolymer**
- **Epoxy composites**
- **Fire resistance**
- **Hydrophilic filler**

REFERENCE

1. Eselev, A. D.; and Bobylev, V. A.; State-of-the-art Production of Epoxy Resins and Adhesive Hardeners in Russia. Adhesives: Sealants; **2006, 7,** 2–8 p.
2. Amirova, L. M.; Ganiev, M. M.; and Amirov, R. R.; Composite Materials Based on Epoxy Oligomers. Kazan: Novoe Znanie Publication; **2002,** 167 p.
3. Kopylov, V. V.; and Novikov, S. N.; Polymer Materials of Reduced Combustibility. Moscow: Khimiya; **1986,** 224 p.
4. Kerber, M. L.; and Vinogradov, V. M.; Polymer Composite Materials: Structure, Properties, and Technology. Ed. Berlin, A. A.; St. Petersburg: Professiya; **2008,** 560 p.

CHAPTER 18

A RESEARCH NOTE ON DEVELOPMENT OF GLUE COMPOSITIONS WITH IMPROVED ADHESION CHARACTERISTICS TO MATERIALS OF DIFFERENT NATURE

V. F. KABLOV, N. A. KEIBAL, S. N. BONDARENKO, D. A. PROVOTOROVA, and G. E. ZAIKOV

CONTENTS

18.1   INTRODUCTION

Today, despite the existence of a large number of adhesives, which differ not only in the composition and properties but also in manufacturing technology, formulations, and intended purpose, the problem of creating new adhesives with a certain set of properties is still relevant. This is due to fact that glue compositions are imposed to ever more high demands related to operation conditions of construction materials and products.

The problem may be solved by applying the targeted modification of a film-forming polymer, which is a base component of any glue composition. Modification is of a priority than creating completely new adhesive formulations. The modification process is more advantageous from both an economic and technological points of view and allows not only to improve the performance of rubber but also to maintain a basic set of properties.

As it is known, there are several methods of film-forming polymer modification. They are physical, chemical, photochemical modification, modification with biologically active systems and combinations of these methods.

Epoxydation, being one of the variants of chemical modification, represents a process of introduction of epoxy groups to a polymer structure that improves properties of this polymer. Materials based on epoxidized polymers show high physical and mechanical characteristics, meet the requirements to the strength and dielectric parameters, and manifest good adhesion to metals, which is achieved due to high adhesive activity of epoxy groups. Thanks to these properties they find application as coatings for metals and plastics, adhesives, mastics, and potting compounds in electrotechnics, microelectronics, and other areas of engineering [1].

Chlorinated natural rubber (CNR) is applied as an additive in glue compositions based on chloroprene rubber that, in its turn, is widely used in industry for gluing of different rubbers together with metals or each other [2]. As an individual, brand glues based on CNR are rarely produced.

Isoprene rubber is an analog of natural rubber, used in the majority of rubber adhesives, but due to its low cohesive strength applied in their formulations much more rarely.

Thereby, the investigations concerned with the development of glue compositions based on these rubbers with improved adhesion characteristics to materials of different nature are of particular interest. It can be achieved by modification [3].

18.2 METHODOLOGY

In this work, we investigated a possibility of chlorinated natural rubber and isoprene rubber epoxydation by means of ozonation with the aim of improving adhesion characteristics of glues based on these rubbers as it is known that epoxy compounds are good film formers in glue compositions and increase the overall viscosity of the last ones.

The introduction of epoxy groups in a polymer structure was carried out by means of ozonation as ozone is highly reactive toward double bonds, aromatic structures, and C–H groups of a macrochain.

The contact time (0.5–2 h) was varied in the ozonation process. The rest of the parameters, which are ozone concentration ($5*10^{-5}$ % vol.) and temperature (23°C), were kept constant. Further, glue compositions based on the ozonized rubbers were prepared.

Glue compositions based on the ozonized CNR were 20 per cent solutions of the rubber in an organic solvent which was ethyl acetate. The compositions based on the isoprene rubber were 5 per cent solutions of the ozonized rubber in petroleum solvent.

The gluing process was conducted at 18–25°C with a double-step deposition of glue and storage of the glued samples under a load of 2 kg for 24 h. Glue bonding of vulcanizates was tested in 24 (± 0.5) h after constructing of joint by method called "shear strength determination" (State Standard 14759-69), in quality of samples, there were used polyisoprene (SKI-3), ethylene propylene (SKEPT-40), butadiene-nitrile (SKN-18), and chloroprene (Neoprene) vulcanized rubbers.

18.3 RESULTS AND DISCUSSION

In ozonation, a partial double-bond breakage in the rubber macromolecules that leads to the macroradicals formation occurs (Figure 18.1). Ozone molecules are attached to the point of the rubber double-bond breakage with the formation of epoxy groups [4]:

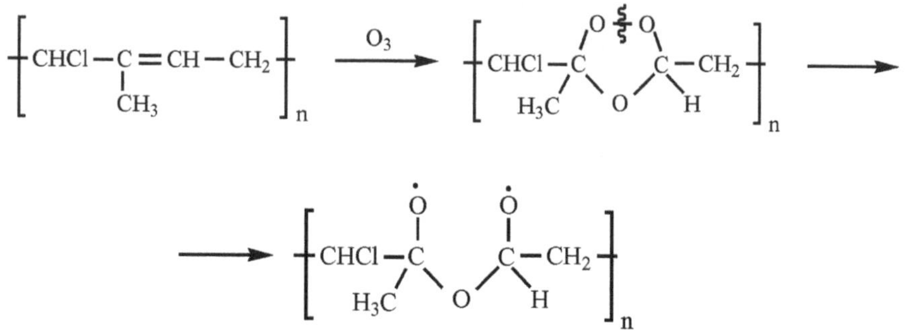

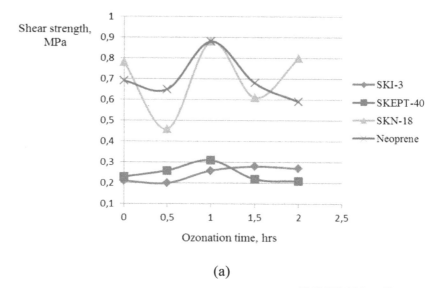

FIGURE 18.1   Reaction of the macroradicals formation (on the example of CNR).

The formed macroradicals are probably interacting with macromolecules of the rubber, which is a substrate material, thereby higher adhesion strength is provided.

At first, the CNR of three brands was investigated; and they are as follows: S-20, CR-10, and CR-20. The results obtained in ozonation of three rubber brands are shown on the Figure (18.2).

(a)

FIGURE 18.2   *(Continued)*

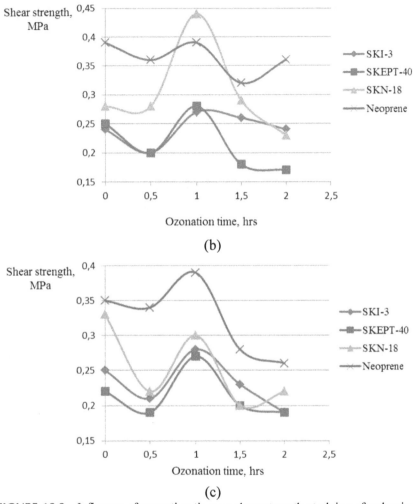

FIGURE 18.2   Influence of ozonation time on shear strength at gluing of vulcanizates with glue compositions based on CNR of the S-20 (a) CR-10, (b) CR-20, and (c) brands accordingly.

A decrease in the adhesion strength at $\tau = 0.5$ h may be connected with preliminary destruction of macromolecules under action of reactive ozone.

From Figure (18.2), we can see that the maximal figures correspond to 1 h ozonation. The adhesion strength for rubbers based on different caoutchoucs increases by 10–40 per cent at that. The results of the shear strength change depending on the rubber brand and adherend type are shown in Figure (18.3).

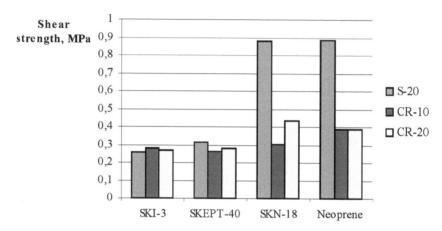

FIGURE 18.3 A change in shear strength for different CNR brands depending on adherent type (ozonation time $\tau = 1$ h).

It should be noticed that the extreme nature of the above-mentioned dependences can be explained by the diffusion nature of the interaction between adhesive and substrate. As it is shown from the figures, with the increasing content of functional groups, the strength began to reduce, having reached of certain limit in the adhesive. In this case, only adhesive molecules have ability to diffusion [5].

Isoprene rubber was also treated with ozone at the same parameters maintained for CNR. The results are shown in Figure (18.4).

The data in Figure (18.4) confirm the ambiguity of the ozonation process. When the contact time is equal to 15 min (as in the case of CNR epoxydation [6]), possible preliminary destruction of the rubber macromolecules takes place, which on the graphs is proved by almost simultaneous reduction in the shear strength values. Concurrently, formation and subsequent growth of macroradicals go on, as evidenced by the increase in adhesive characteristics at ozonation time 0.5 and 1 h. Here, the shear strength at gluing of different vulcanized rubbers increases by 10–70 per cent on average, and then it starts to reduce again.

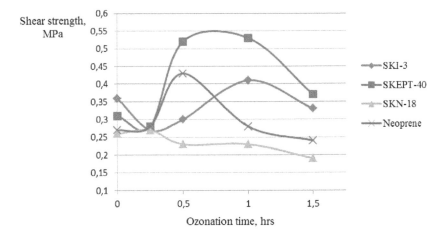

FIGURE 18.4 Influence of ozonation time on adhesive strength for the compositions based on isoprene rubber.

With further increase in ozonation time, the values of adhesion strength decline. That is apparently related to saturation of the polymer chain with epoxy groups and decrease in mobility of the macromolecules, and, consequently, the degree of interaction of the substrate with the adhesive composition as well as with destruction of the polymer chains.

18.4 CONCLUSIONS

Thus, ozonation can be applied as an effective method of enhancing adhesion properties of rubbers at the modification of film-forming polymers that are a main component in glues. By changing one of the parameters during the ozonation process, we can obtain such content of epoxy groups at which the characteristics of adhesion strength will be maximal.

KEYWORDS

- **Glue compositions**
- **Gluing**
- **Modification**
- **Ozonation**
- **Unsaturated rubbers**
- **Vulcanizates**

BIBLIOGRAPHY

1. Solovyev, M. M.; Local Dynamics of Oligobutadienes of Different Microstructure and Their Modification Products: Thesis of PhD in Chemistry Sciences: 02.00.06. Yaroslavl: Solovyev Mikhail Mikhailovich; **2009,** 201 p.
2. Dontsov, A. A.; Lozovick, G. Ya.; and Novitskaya, S. P.; Chlorinated Polymers. Moscow: Khymiya; **1979,** 232 p.
3. Kablov, V. F.; Bondarenko, S. N.; and Keibal, N. A.; Modification of Elastic Glue Compositions and Coatings with Element Containing Adhesion Promoters: Monograph. Volgograd: IUNL VSTU; **2010,** 238 p.
4. Zaikov, G. E.; Why do polymers age. *Soros Educ. J.* **2000,** *6(12),* 52.
5. Berlin, A. A.; and Basin, V. E.; Basics of Polymer Adhesion. Moscow: Khymiya; **1969,** 320 p.
6. Keibal, N. A.; Bondarenko, S. N.; Kablov, V. F.; and Provotorova, D. A.; Ozonation of chlorinated natural rubber and studying its adhesion characteristics. In: Rubber: Types, Properties and Uses. Ed. Popa, Gabriel A.; New York: Nova Publishers; **2012,** 275–280 p.

CHAPTER 19

UPDATE ON BASIC MATHEMATICAL RELATIONS WITH PARTICULAR REFERENCE ON INDUSTRIAL DRYING OF LEATHER

A. K. HAGHI and G. E. ZAIKOV

CONTENTS

19.1　INTRODUCTION

The manufacturing of new-generation synthetic leathers involves the extraction of the filling polymer from the polymer-matrix system with an organic solvent and the removal of the solvent from the highly porous material. In this paper, a mathematical model of synthetic leather drying for removing the organic solvent is proposed [1–10]. The model proposed adequately describes the real processes. To improve the accuracy of calculated moisture distributions a velocity correction factor (VCF) introduced into the calculations. The VCF reflects the fact that some of the air flowing through the bed does not participate very effectively in drying, since it is channeled into low-density areas of the inhomogeneous bed. The present chapter discusses the results of experiments to test the deductions that increased rates of drying and better agreement between predicted and experimental moisture distributions in the drying bed can be obtained by using higher air velocities [11–25].

The present work focuses on reviewing convective heat and mass transfer equations in the industrial leather drying process with particular reference to VCF.

19.2　CLASSICAL MATHEMATICAL RELATIONS

The theoretical model proposed in this article is based on fundamental equations to describe the simultaneous heat and mass transfer in porous media. It is possible to assume the existence of a thermodynamic quasi equilibrium state, where the temperatures of gaseous, liquid, and solid phases are equal, i.e.,

$$T_S = T_L = T_G = T \tag{19.1}$$

Liquid Mass Balance:

$$\frac{\partial(\varepsilon_L \rho_L)}{\partial t} + \nabla(\rho_L \vec{u}_L) + \dot{m} = 0 \tag{19.2}$$

-Water Vapor Mass Balance:

$$\frac{\partial\left[(\varepsilon-\varepsilon_L)X_V\rho_G\right]}{\partial t}+\nabla\left(X_V\rho_G\vec{u}_G+\vec{J}_V\right)-\dot{m}=0 \qquad (19.3)$$

$$\vec{J}_V=-\rho_G\left(\varepsilon-\varepsilon_L\right)D_{EFF}\nabla X_V \qquad (19.4)$$

-Air Mass Balance:

$$\frac{\partial\left((\varepsilon-\varepsilon_L)X_A\rho_G\right)}{\partial t}+\nabla\left(X_A\rho_G\vec{u}_G-\vec{J}_V\right)=0 \qquad (19.5)$$

-Liquid Momentum Equation (Darcy's Law):

$$\vec{u}_L=-\left(\frac{\alpha_G}{\mu_G}\right)\nabla(P_G) \qquad (19.6)$$

-Thermal Balance:
The thermal balance is governed by Eq. (19.7).

$$\frac{\partial\left\{\left[\rho_s C_{Ps}+(\varepsilon-\varepsilon_L)\rho_G\left(X_V C_{Pv}+X_A C_{PA}\right)+\varepsilon_L\rho_L C_{PL}\right]T\right\}}{\partial t}-\nabla\left(k_E\nabla T\right)+$$
$$\nabla\left[\left(\rho_L\vec{u}_L C_{PL}+\rho_G\vec{u}_G\left(X_V C_{Pv}+X_A C_{PA}\right)\right)T\right]+(\varepsilon-\varepsilon_L)\frac{\partial P_G}{\partial t}+\dot{m}\Delta H_V=0 \qquad (19.7)$$

-Thermodynamic Equilibrium-Vapor mass Fraction:
In order to attain thermal equilibrium between the liquid and vapor phase, the vapor mass fraction should be such that the partial pressure of the vapor (P'_V) should be equal to its saturation pressure (P_{VS}) at temperature of the mixture. Therefore, thermodynamic relations can obtain the concentration of vapor in the air/vapor mixture inside the pores. According to Dalton's Law of Additive Pressure applied to the air/vapor mixture, one can show that [26–37]:

$$\rho_G=\rho_V+\rho_A \qquad (19.8)$$

$$X_V=\frac{\rho_V}{\rho_G} \qquad (19.9)$$

$$\rho_V = \frac{P'}{R_V T} \tag{19.10}$$

$$\rho_A = \frac{(P_G - P'_V)}{R_A T} \tag{19.11}$$

Combining Eqs. (19.8–19.11), one can obtain:

$$X_V = \frac{1}{1 + \left(\dfrac{P_G R_V}{P'_V R_A}\right) - \left(\dfrac{R_V}{R_A}\right)} \tag{19.12}$$

-Mass Rate of Evaporation:

The mass rate of evaporation was obtained in two different ways, as follows [38–49]:

First of all, the mass rate of evaporation $\dot{m}$ was expressed explicitly by taking it from the water vapor mass balance Eq. (19.2), since vapor concentration is given by Eq. (19.12).

$$\dot{m} = \frac{\partial\left[(\varepsilon - \varepsilon_L) X_V \rho_G\right]}{\partial t} + \nabla\left(X_V \rho_G \vec{u}_G + \vec{J}_V\right) \tag{19.13}$$

Secondly, an equation to compute the mass rate of evaporation can be derived with a combination of the liquid mass balance Eq. (19.1) with a first-order-Arthenius type equation. From the general kinetic equation:

$$\frac{\partial \alpha}{\partial t} = -kf(\alpha) \tag{19.14}$$

$$k = A\exp\left(-\frac{E}{RT_{SUR}}\right) \tag{19.15}$$

$$\alpha = 1 - \frac{\varepsilon_L(t)}{\varepsilon_0} \tag{19.16}$$

-Drying Kinetics Mechanism Coupling:

Using thermodynamic relations, according to Amagat's law of additive volumes, under the same absolute pressure [50–72],

$$m_V = \frac{V_V P_G}{R_V T} \qquad (19.17)$$

$$m_A = \frac{V_A P_G}{R_A T} \qquad (19.18)$$

$$m_V = X_V m_T \qquad (19.19)$$

$$m_T = m_V + m_A \qquad (19.20)$$

$$V_G = V_V + V_A \qquad (19.21)$$

$$V_G = (\varepsilon - \varepsilon_L) V_S \qquad (19.22)$$

Solving the set of Algebraic Eqs. (19.17–19.22), one can obtain the vapor-air mixture density:

$$\rho_G = \frac{(m_V + m_A)}{V_G} \qquad (19.23)$$

$$\rho_V = \frac{m_V}{V_G} \qquad (19.24)$$

$$\rho_A = \frac{m_A}{V_G} \qquad (19.25)$$

-Equivalent Thermal Conductivity [73–85]:

It is necessary to determine the equivalent value of the thermal conductivity of the material as a whole, since no phase separation was considered in the overall energy equation [85–97]. The equation, we can propose now which may be used to achieve the equivalent thermal conductivity of materials K_E, composed of a continued medium with a uniform disperse phase. It is expressed as follows in Eq. (19.26).

$$K_E = \cfrac{\left[k_S + \varepsilon_L k_L \left(\cfrac{3k_S}{2k_S + k_L} \right) + k_G (\varepsilon - \varepsilon_L) \left(\cfrac{3k_S}{2k_S + k_G} \right) \right]}{\left[1 + \varepsilon_L \left(\cfrac{3k_S}{2k_S + k_L} \right) + (\varepsilon - \varepsilon_L) \left(\cfrac{3k_S}{2k_S + k_G} \right) \right]} \tag{19.26}$$

$$k_G = X_V k_V + X_A k_A \tag{19.27}$$

-Effective Diffusion Coefficient Equation:
The binary bulk diffusivity D_{AV} of air-water vapor mixture is given by:

$$D_{AV} = (2.20)(10^{-5}) \left(\frac{P_{ATM}}{P_G} \right) \left(\frac{T_{REF}}{273.15} \right)^{1.75} \tag{19.28}$$

Factor α_F can be used to account for closed pores resulting from different nature of the solid, which would increase gas outflow resistance, so the equation of effective diffusion coefficient D_{EFF} for fiber drying is:

$$D_{EFF} = \alpha_F D_{AV} \tag{19.29}$$

The convective heat-transfer coefficient can be expressed as:

$$h = Nu_\delta \left(\frac{k}{\delta} \right) \tag{19.30}$$

The convective mass transfer coefficient is:

$$h_M = \left(\frac{h}{C_{PG}} \right) \left(\frac{Pr}{Sc} \right)^{2/3} \tag{19.31}$$

$$Pr = \frac{C_{PG} \mu_G}{k_G} \tag{19.32}$$

$$Sc = \frac{\mu_G}{\rho_G D_{AV}} \tag{19.33}$$

The deriving force determining the rate of mass transfer inside the fiber is the difference between the relative humidities of the air in the pores and the fiber. The rate of moisture exchange is assumed to be proportional to the relative humidity difference in this study.

The heat-transfer coefficient between external air and fibers surface can be obtained by:

$$h = Nu_\delta \left(\frac{k}{\delta} \right)$$

The mass transfer coefficient was calculated using the analogy between heat-transfer and mass transfer as $h_M = \left(\frac{h}{C_{PG}} \right) \left(\frac{\text{Pr}}{\text{Sc}} \right)^{2/3}$. The convective heat and mass transfer coefficients at the surface are important parameters in drying processes; they are functions of velocity and physical properties of the drying medium.

19.3 A KINETIC MODEL

Describing kinetic model of the moisture transfer during drying as follows:

$$-\frac{dX}{dt} = k(X - X_e) \tag{19.34}$$

Where, X is the material moisture content (dry basis) during drying (kg water/kg dry solids), X_e is the equilibrium moisture content of dehydrated material (kg water/kg dry solids), k is the drying rate (min^{-1}), and t is the time of drying (min). The drying rate is determined as the slope of the falling rate-drying curve. At zero time, the moisture content (dry basis) of the dry material X (kg water/kg dry solids) is equal to X_i, and Eq. (19.34) is integrated to give the following expression:

$$X = X_e - (X_e - X_i)e^{-kt} \tag{19.35}$$

Using above equation Moisture Ratio can be defined as follows:

$$\frac{X - X_e}{X_i - X_e} = e^{-kt} \tag{19.36}$$

This is the Lewis's formula introduced in 1921. But using experimental data of leather drying it seemed that there was an error in curve fitting of e^{-at}.

The experimental moisture content data were non-dimensionlized using the equation:

$$MR = \frac{X - X_e}{X_i - X_e} \qquad (19.37)$$

where MR is the moisture ratio. For the analysis it was assumed that the equilibrium moisture content, X_e was equal to zero.

Selected drying models, detailed in Table (19.1), were fitted to the drying curves (MR versus time), and the equation parameters determined using non-linear least squares regression analysis, as shown in Table (19.2).

TABLE 19.1   Drying models fitted to experimental data

Model	Mathematical expression
Lewis (1921)	$MR = \exp(-at)$
Page (1949)	$MR = \exp(-at^b)$
Henderson and Pabis (1961)	$MR = a\exp(-bt)$
Quadratic function	$MR = a + bt + ct^2$
Logarithmic (Yaldiz and Eterkin, 2001)	$MR = a\exp(-bt) + c$
3rd Degree polynomial	$MR = a + bt + ct^2 + dt^3$
Rational function	$MR = \dfrac{a + bt}{1 + ct + dt^2}$
Gaussian model	$MR = a\exp(\dfrac{-(t-b)^2}{2c^2})$
Present model	$MR = a\exp(-bt^c) + dt^2 + et + f$

TABLE 19.2 Estimated values of coefficients and statistical analysis for the drying models

Model	Constants	T = 50	T = 65	T = 80
Lewis	a	0.08498756	0.1842355	0.29379817
	S	0.0551863	0.0739872	0.0874382
	r	0.9828561	0.9717071	0.9587434
Page	a	0.033576322	0.076535988	0.14847589
	b	1.3586728	1.4803604	1.5155253
	S	0.0145866	0.0242914	0.0548030
	r	0.9988528	0.9972042	0.9856112
Henderson	a	1.1581161	1.2871764	1.4922171
	b	0.098218605	0.23327801	0.42348755
	S	0.0336756	0.0305064	0.0186881
	r	0.9938704	0.9955870	0.9983375
Logarithmic	a	1.246574	1.3051319	1.5060514
	b	0.069812589	0.1847816	0.43995186
	c	−0.15769402	−0.093918118	0.011449769
	S	0.0091395	0.0117237	0.0188223
	r	0.9995659	0.9993995	0.9985010
Quadratic function	a	1.0441166	1.1058544	1.1259588
	b	−0.068310663	−0.16107942	−0.25732004
	c	0.0011451149	0.0059365371	0.014678241
	S	0.0093261	0.0208566	0.0673518
	r	0.9995480	0.9980984	0.9806334
3rd degree polynomial	a	1.065983	1.1670135	1.3629748
	b	−0.076140508	−0.20070291	−0.45309695
	c	0.0017663191	0.011932525	0.053746805

	d	$-1.335923\,e-005$	-0.0002498328	-0.0021704758
	s	0.0061268	0.0122273	0.0320439
	r	0.9998122	0.9994013	0.9961941
Rational function	a	1.0578859	1.192437	1.9302135
	b	−0.034944627	−0.083776453	−0.16891461
	c	0.03197939	0.11153663	0.72602847
	d	0.0020339684	0.01062793	0.040207428
	s	0.0074582	0.0128250	0.0105552
	r	0.9997216	0.9993413	0.9995877
Gaussian model	a	1.6081505	2.3960741	268.28939
	b	−14.535231	−9.3358707	−27.36335
	c	15.612089	7.7188252	8.4574493
	s	0.0104355	0.0158495	0.0251066
	r	0.9994340	0.9989023	0.9973314
Present model	a	0.77015136	2.2899114	4.2572457
	b	0.073835826	0.58912095	1.4688178
	c	0.85093985	0.21252159	0.39672164
	d	0.00068710356	0.0035759092	0.0019698297
	e	−0.037543605	−0.094581302	−0.03351435
	f	0.3191907	−0.18402789	0.04912732
	s	0.0061386	0.0066831	0.0092957
	r	0.9998259	0.9998537	0.9997716

The experimental results for the drying of leather are given in Figure (19.1). Fitting curves for two sample models (Lewis model and present model) and temperature of 80°C are given in Figures (19.2) and (19.3). Two criteria were adopted to evaluate the goodness of fit of each model, the Correlation Coefficient (r) and the Standard Error (S). The standard error of the estimate is defined as follows:

$$S = \sqrt{\frac{\sum\limits_{i=i}^{n_{points}} (MR_{exp,i} - MR_{Pred,i})^2}{n_{points} - n_{param}}} \qquad (19.38)$$

Where $MR_{exp,i}$ is the measured value at point i, and $MR_{Pred,i}$ is the predicted value at that point, and n_{param} is the number of parameters in the particular model (so that the denominator is the number of degrees of freedom).

To explain the meaning of correlation coefficient, we must define some terms used as follow:

$$S_t = \sum\limits_{i=1}^{n_{points}} (\bar{y} - MR_{exp,i})^2 \qquad (19.39)$$

Where, the average of the data points ($\bar{y}$) is simply given by

$$\bar{y} = \frac{1}{n_{points}} \sum\limits_{i=1}^{n_{points}} MR_{exp,i} \qquad (19.40)$$

The quantity S_t considers the spread around a constant line (the mean) as opposed to the spread around the regression model. This is the uncertainty of the dependent variable prior to regression. We also define the deviation from the fitting curve as:

$$S_r = \sum\limits_{i=1}^{n_{points}} (MR_{exp,i} - MR_{pred,i})^2 \qquad (19.41)$$

Note the similarity of this expression to the standard error of the estimate given above; this quantity likewise measures the spread of the points around the fitting function. In view of the above, the improvement (or error reduction) due to describing the data in terms of a regression model can be quantified by subtracting the two quantities. Because the magnitude of the quantity is dependent on the scale of the data, this difference is normalized to yield

$$r = \sqrt{\frac{S_t - S_r}{S_t}} \qquad (19.42)$$

Where, r is defined as the correlation coefficient. As the regression model better describes the data, the correlation coefficient will approach unity. For a perfect fit, the standard error of the estimate will approach $S = 0$ and the correlation coefficient will approach $r = 1$.

The standard error and correlation coefficient values of all models are given in Figures (19.4) and (19.5).

19.4   DISCUSSION OF RESULTS

Synthetic leathers are materials with much varied physical properties. As a consequence, even though a lot of research of simulation of drying of porous media has been carried out, the complete validation of these models are very difficult. The drying mechanisms might strongly influenced by parameters such as permeability and effective diffusion coefficients. The unknown effective diffusion coefficient of vapor for fibers under different temperatures may be determined by adjustment of the model's theoretical alpha correction factor and experimental data. The mathematical model can be used to predict the effects of many parameters on the temperature variation of the fibers. These parameters include the operation conditions of the dryer, such as the initial moisture content of the fibers, heat, and mass transfer coefficients, drying air moisture content, and dryer air temperature. From Figures (19.6–19.11), it can be observed that the shapes of the experimental and calculated curves are somewhat different. It can be seen that as the actual air velocity used in this experiment increases, the value of VCF necessary to achieve reasonable correspondence between calculation and experiment becomes closer to unity; a smaller correction to air velocity is required in the calculations as the actual air velocity increases. This appears to confirm the fact that the loss in drying efficiency caused by bed inhomogeneity tends to be reduced as air velocity increases. Figure (19.1) shows a typical heat distribution during convective drying. Table (19.3) relates the VCF to the values of air velocity actually used in the experiments. It is evident from the table that the results show a steady improvement in agreement between experiment and calculation (as indicated by increase in VCF) for air velocities up to 1.59 m/s, above which to be no further improvement with increased flow.

TABLE 19.3   Variation of VCF with air velocity

Air velocity, (m/s)	0.75	0.89	0.95	1.59	2.10	2.59
VCF used	0.39	0.44	0.47	0.62	0.62	0.61

In this work, the analytical model has been applied to several drying experiments. The results of the experiments and corresponding calculated distributions are shown in Figures (19.6–19.11). It is apparent from the curves that the calculated distribution is in reasonable agreement with the corresponding experimental one. In view of the above, it can be clearly observed that the shapes of experimental and calculated curves are somewhat similar.

19.5   SUMMARY

It is observed that the high air velocities will reduce the thickness of the stationary gas film on the surface of the solid and hence, increase the heat and mass transfer coefficients. In practical designing of dryers, it is found to be more reliable to consider heat-transfer rates than mass transfer rates, as the latter are a function of surface temperature of the wet solid, which is difficult to determine and cannot, in practice, be assumed to be that of the wet-bulb temperature of the air with an adequate degree of accuracy. In the model presented in this paper, a simple method of predicting moisture distributions leads to prediction of drying times more rapid than those measured in experiments. From this point of view, the drying reveals many aspects, which are not normally observed or measured and which may be of value in some application.

The derivation of the drying curves is an example. It is clear from the experiments over a range of air velocities that it is not possible to make accurate predictions and have the experimental curves coincide at all points with the predicted distributions simply by introducing a VCF into the calculations. This suggest that a close agreement between calculated and experimental curves over the entire drying period could be obtained by using a large value of VCF in the initial stages of drying and progressively decreasing it as drying proceeds.

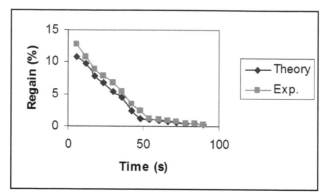

FIGURE 19.1 Comparison of the theoretical and experimental. Distribution at air velocity of 0.75 m/s and VCF = 0.39.

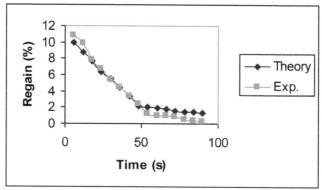

FIGURE 19.2 Comparison of the theoretical and experimental. Distribution at air velocity of 0.89 m/s and VCF = 0.44.

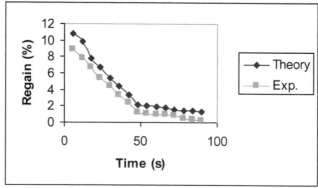

FIGURE 19.3 Comparison of the theoretical and experimental. Distribution at air velocity of 0.95 m/s and VCF = 0.47.

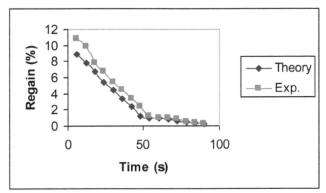

FIGURE 19.4 Comparison of the theoretical and experimental. Distribution at air velocity of 1.59 m/s and VCF = 0.62.

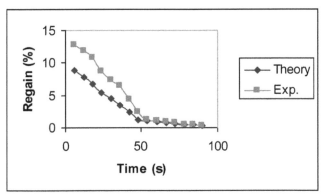

FIGURE 19.5 Comparison of the theoretical and experimental. Distribution at air velocity of 2.10 m/s and VCF = 0.62.

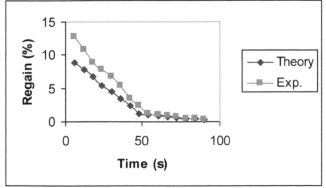

FIGURE 19.6 Comparison of the theoretical and experimental. Distribution at air velocity of 2.59 m/s and VCF = 0.61.

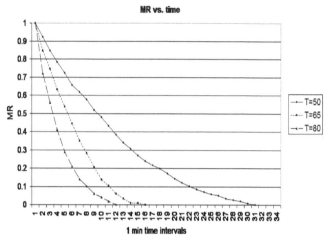

FIGURE 19.7　Moisture ratio vs. time.

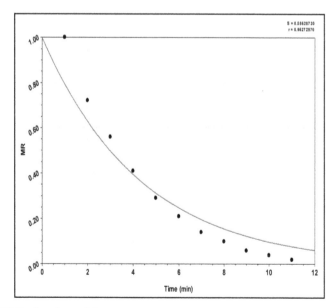

FIGURE 19.8　Lewis model.

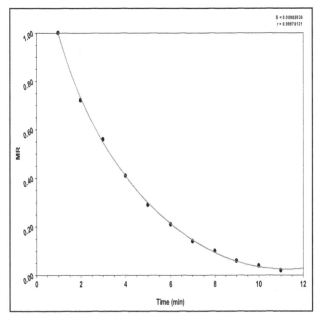

FIGURE 19.9   Present model.

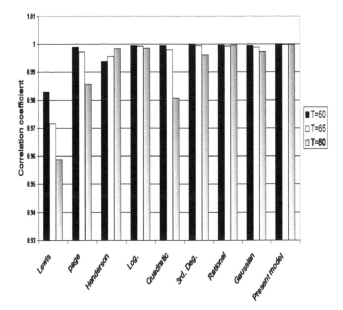

FIGURE 19.10   Correlation coefficient of all models.

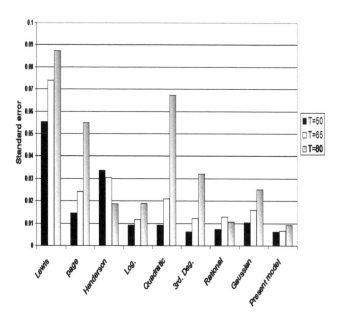

FIGURE 19.11   Standard error of all models.

NOMENCLATURE

A	area
a,b,c	constants
c_p	constant pressure specific heat
C_A	moisture content of air in fabric pores
C_a	water-vapor concentration in the air filling the inter-fiber void space
C_e	moisture content of extent air
C_F	moisture content of fibers in a fabric
C_f	water-vapor concentration in the fibers of the fabric (kg)
C_p	specific heat
D	diffusion coefficient
D_b	bound water conductivity
D_{eff}	effective diffusivity

E_d	activation energy of movement of bound water
h	enthalpy (J/kg)
h_e	heat-transfer coefficient
h_m	mass transfer coefficient
	enthalpy of vaporization (J/kg)
Δh_{vap}	latent heat of evaporation
J	species diffusion flux
J_L	free water flux
K	permeability
K_0	single-phase permeability of porous material
K_r	relative permeability
k	thermal conductivity
k_{eff}	effective thermal conductivity
m	ratio of diffusion coefficients of air and water vapor
m'	mass source per unit volume
$\dot{m}$	evaporation rate, mass transfer rate
M	molecular weight
p	pressure
P_c	capillary pressure
P_s	saturation pressure
q	convective heat-transfer rate
Q	enthalpy of desorption from solid phase
r	radius
R	gas constant, Fiber regain
S	pore saturation
t	Time
T	Temperature
T_e	external air temperature
U	moisture content

GREEK SYMBOLS

γ	pore volume density function
λ	latent heat of evaporation
λ_{eff}	effective thermal conductivity
μ	Viscosity
ν	fluid velocity
ρ	Density
σ	surface tension
τ	tortuosity factor of capillary paths
ψ	relative humidity
ω	averaging volume
ε	volume fraction (of quantity)

SUBSCRIPTS

0	Initial
c	capillary, critical
eq	Equilibrium
g	Gas
ir	Irreducible
ms	maximum sorptive
v	Vapor
w	Water
	Liquid phase
	Gas phase
	Solid phase
bl	Bound liquid
ds	Dry solid
lv	Liquid-to-vapor
ls	Liquid-to-solid

sat	Saturation
sv	Solid-to-vapor
v	Vapor

SUPERSCRIPTS

g	intrinsic average over the gaseous phase
l	intrinsic average over the liquid phase
*	vapor saturated
-	average value

KEYWORDS

- **Classical note**
- **Mathematical relations**
- **Leather drying**

REFERENCE

1. Armor, J.; and Cannon, J.; Fluid flow through woven screens. *AICHE J.* **1968,** *14(3),* 415–420.
2. ASTMD737-75, Standard Test Methods for Air Permeability of Textile Fabrics.
3. ASTME96-95, Water Vapour Transmission of Materials.
4. Arnod, G.; and Fohr, J. P.; Slow drying simulation in thick layears of granular products. *Int. J. Heat Mass Transfer.* **1988,** *31(12),* 2517–2562.
5. Azizi, S.; Moyne, C.; and Degiovanni, A.; Approche experimentale et theorique de la conductivite thermique des milieux poreux humides. *Int. J. Heat Mass Transfer.* **1988,** *31(11),* 2305–2317.
6. Backer, S.; The relationship between the structural geometry of a textile fabric and its physical properties Part IV: interstice geometry and air permeability. *Textile Res. J.* **1951,** *21,* 703–714.
7. Barnes, J.; and Holcombe, B.; Moisture sorption and transport in clothing during wear. *Textile Res. J.* **1996,** *66(12),* 777–786.
8. Bartles, V. T.; Survey on the Moisture Transport Properties of Foul Weather Protective Textiles at Temperatures around and Below the Freezing Point. Technical Report No. 11674, Boennigheim, Germany: Hohenstein Institute of Clothing Physiology; **2001.**
9. Bears, J.; "Dynamics of Fluids in Porous Media." New York: Elsevier; **1972.**
10. Black, W. Z.; and Hartley, J. G.; Thermodynamics. New York: Harper & Row; **1985.**

11. BS 4407. Quantitative Analysis of Fiber Mixtures. **1997**.

12. BS 7209. Specification for Water Vapour Permeable Apparel Fabrics. **1990**.

13. CAN2-4.2-M77. Method of Test for Resistance of Materials to Water Vapour Diffusion (Control Dish Method). **1977**.

14. CGSB-4.2 No. 49-M91. Resistance of Materials to Water Vapour Diffusion.

15. Chen, C. S.; and Johnson, W. H.; Kinetics of moisture movement in hygroscopic materials. In: Theoretical Considerations of Drying Phenomenon. Trans. ASAE.; **1969**, *12*, 109–113.

16. Chen, P.; and Pei, D.; A mathematical model of drying process. *Int. J. Heat Mass Transfer.* **1988**, *31(12)*, 2517–2562.

17. Chen, P.; and Schmidt, P. S.; An integral model for drying of hygroscopic and non-hygroscopic materials with dielectric heating. *Drying Technol.* **1990**, *8(5)*, 907–930.

18. Chen, P.; and Schmidt, P. S.; A model for drying of flow-through beds of granular products with dielectric heating. In: Transport Phenomena in Materials Processing. ASME, New York: American Society of Mechanical Engineers, Heat-transfer Division, (Publication) HDT; **1990**, *146*, 121–127.

19. Davis, A.; and James, D.; Slow flow through a model fibrous porous medium. *Int. J. Multiphase Flow.* **1996**, *22*, 969–989.

20. Dietl, C.; George, O. P.; and Bansal, N. K.; Modeling of diffusion in capillary porous materials during the drying process. *Drying Technol.* **1995**, *13(1 & 2)*, 267–293.

21. Ea, J. Y.; Water Vapour Transfer in Breathable Fabrics for Clothing. PhD Thesis, University of Leeds, **1988**.

22. Flory, P. J.; Statistical Mechanics of Chain Molecules. New York: Inter Science Publication; **1969**.

23. Francis, N. D.; and Wepfer, W. J.; Jet impeingement drying of a moist porous solid. *Int. J. Heat Mass Transfer.* **1996**, *39(9)*, 1911–1923.

24. Gerald, C. F.; and Wheatley, P. O.; Applied Numerical Analysis. 4th Ed. Reading, MA: Addison-Wesley; **1989**.

25. Ghali, K.; Jones, B.; and Tracy, E.; Experimental techniques for measuring parameters describing wetting and wicking in fabrics. *Text. Res. J.* **1994**, 106–111.

26. Gibson, P.; Elsaiid, A.; Kendrick, C. E.; Rivin, D.; and Charmchi, M.; A test method to determine the relative humidity dependence of the air permeability of textile materials. *J. Test. Eval.* **1997**, *25(4)*, 416–423.

27. Gibson, P.; and Charmchi, M.; The use of volume-averaging techniques to predict temperature transients due to water vapour sorption in hygroscopic porous polymer materials. *J. Appl. Polym. Sci.* **1997**, *64*, 493–505.

28. Ghali, K.; Jones, B.; and Tracy, E.; Modeling heat and mass transfer in fabrics. *Int. J. Heat Mass Transfer.* **1995**, *38(1)*, 13–21.

29. Gennes, P. G.; Scaling Concepts in Polymer Physics. 3rd Ed. Ithaca, NY: Cornell University Press; **1988**.

30. Givoni, B.; and Goldman, R. F.; Predicting metabolic energy cost. *J. Appl. Physiol.* **1971**, *30(3)*, 429–433.

31. Green, J. H.; "An Introduction to Human Physiology." J. Comyn, Ed. London: Elsevier Applied Science Publishers; **1985**.

32. Greenkorn, R. A.; "Flow Phenomena in Porous Media." New York: Marcel Dekker; **1984**.

33. Hadley, G. R.; Numerical modeling of the drying of porous materials. In: Proceedings of The Fourth International Drying Symposium. **1984,** *1,* 151–158.

34. Haghi, A. K.; Moisture permeation of clothing. *JTAC.* **2004,** *76,* 1035–1055.

35. Haghi, A. K.; Thermal analysis of drying process. *JTAC.* **2003,** *74,* 827–842.

36. Haghi, A. K.; Some Aspects of Microwave Drying. The Annals of Stefan cel Mare University. Year VII, **2000,** *14,* 22–25.

37. Haghi, A. K.; A Thermal Imaging Technique for Measuring Transient Temperature Field- an Experimental Approach. The Annals of Stefan cel Mare University. Year VI, **2000,** *12,* 73–76.

38. Haghi, A. K.; Experimental investigations on drying of porous media using infrared radiation. *Acta Polytech.* **2001,** *41(1),* 55–57.

39. Haghi, A. K.; A mathematical model of the drying process. *Acta Polytech.* **2001,** *41(3),* 20–23.

40. Haghi, A. K.; Simultaneous moisture and heat-transfer in porous system. *J. Comput. Appl. Mech.* **2001,** *2(2),* 195–204.

41. Haghi, A. K.; A detailed study on moisture sorption of hygroscopic fiber. *J. Theor. Appl. Mech.* **2002,** *32(2),* 47–62.

42. Haghi, A. K.; A mathematical approach for evaluation of surface topography parameters. *Acta Polytech.* **2002,** *42(2),* 35–40.

43. Haghi A. K.; Mechanism of heat and mass transfer in moist porous materials. *J. Technol.* **2002,** *35(F),* 1–16.

44. Haghi, A. K.; A study of drying process. *H.J.I.C.* **2002,** *30,* 261–269.

45. Haghi, A. K.; Experimental evaluation of the microwave drying of natural silk. *J. Theor. Appl. Mech.* **2003,** *33,* 83–94.

46. Haghi, A. K.; Mahfouzi, K.; and Mohammadi, K.; The effects of microwave irradiations on natural silk. *JUCTM.* **2002,** *38,* 85–96.

47. Haghi, A. K.; The diffusion of heat and moisture through textiles. *Int. J. Appl. Mech. Eng.* **2003,** *8(2),* 233–243.

48. Haghi, A. K.; and Rondot, D.; Heat and mass transfer of leather in the drying process. *IJC Chem. Eng.* **2004,** *23,* 25–34.

49. Haghi, A. K.; Heat and mass transport through moist porous materials. *14th Int. Symp. Transport Phenom. Proc.* 209–214, July 6–9, **2003,** Indonesia.

50. Hartley, J. G.; Coupled heat and moisture transfer in soils: a review. *Adv. Drying.* *2003, 4,* 199–248.

51. Higdon, J.; and Ford, G.; Permeability of three-dimensional models of fibrous porous media. *J. Fluid Mechan.* **1996,** *308,* 341–361.

52. Hong, K.; Hollies, N. R. S.; and Spivak, S. M.; Dynamic moisture vapour transfer through textiles, Part I: clothing hygrometry and the influence of fiber type. *Text. Res. J.* **1988,** *58(12),* 697–706.

52. Hsieh, Y. L.; Yu, B.; and Hartzell, M.; Liquid wetting transport and retention properties of fibrous assemblies, Part II: water wetting and retention of 100 per cent and blended woven fabrics. *Text. Res. J.* **1992,** *62(12),* 697–704.

53. Huh, C.; and Scriven, L. E.; Hydrodynamic model of steady movement of a solid-liquid-fluid contact line. *J. Coll. Int. Sci.* **1971,** *35,* 85–101.

54. Incropera, F. P.; and Dewitt, D. P.; Fundamentals of Heat and Mass Transfer. 2nd Ed. New York: Wiley; **1985.**

55. ISO 11092. Measurement of Thermal and Water-Vapour Resistance under Steady-State Conditions (Sweating Guarded-Hotplate Test). **1993**.
56. Ito, H.; and Muraoka, Y.; Water transport along textile fibers as measured by an electrical capacitance technique. *Text. Res. J.* **1993**, *63(7)*, 414–420.
57. Jackson, J.; and James, D.; The permeability of fibrous porous media. *Can. J. Chem. Eng.* **1986**, *64*, 364–374.
58. Jacquin, C. H.; and Legait, B.; Influence of capillarity and viscosity during spontaneous imbibition in porous media and capillaries. *Phys. Chem. Hydro.* **1984**, *5*, 307–319.
59. Jirsak, O.; Gok, T.; Ozipek, B.; and Pau, N.; Comparing dynamic and static methods for measuring thermal conductive properties of textiles. *Text. Res. J.* **1998**, *68(1)*, 47–56.
60. Kaviany, M.; "Principle of Heat-transfer in Porous Media." New York: Springer; **1991**.
61. Keey, R. B.; "Drying: Principles and Practice." Pergamon: Oxford; **1975**.
62. Keey, R. B.; "Introduction to Industrial Drying Operations." Pergamon: Oxford; **1978**.
63. Keey, R. B.; The drying of textiles. *Rev. Prog. Color.* **1993**, *23*, 57–72.
64. Kulichenko, A.; and Langenhove, L.; The resistance to flow transmission of porous materials. *J. Text. Inst.* **1992**, *83(1)*, 127–132.
65. Kyan, C.; Wasan, D.; and Kintner, R.; Flow of single-phase fluid through fibrous beds. *Ind. Eng. Chem. Fundam.* **1970**, *9(4)*, 596–603.
66. Le, C. V.; and Ly, N. G.; Heat and moisture transfer in textile assemblies, Part I: steaming of wool, cotton, nylon, and polyester fabric beds. *Text. Res. J.* **1995**, *65(4)*, 203–212.
67. Le, C. V.; Tester, D. H.; and Buckenham, P.; Heat and moisture transfer in textile assemblies, Part II: steaming of blended and interleaved fabric-wrapper assemblies. *Text. Res. J.* **1995**, *65(5)*, 265–272.
68. Lee, H. S.; Carr, W. W.; Becckham, H. W.; and Wepfer, W. J.; Factors influencing the air flow through unbacked tufted carpet. *Text. Res. J.* **2000**, *70*, 876–885.
69. Lee, H. S.; Study of the Industrial Through-Air Drying Process for Tufted Carpet. Doctoral Thesis, Atlanta, GA: Georgia Institute of Technology; **2000**.
70. Luikov, A. V.; Heat and Mass Transfer in Capillary Porous Bodies. Oxford: Pergamon Press; **1966**.
71. Luikov, A. V.; Systems of differential equations of heat and mass transfer in capillary porous bodies. *Int. J. Heat Mass Transfer.* **1975**, *18*, 1–14.
72. Metrax, A. C.; and Meredith R. J.; Industrial Microwave Heating. London, England: Peter Peregrinus Ltd; **1983**.
73. Mitchell, D. R.; Tao, Y.; and Besant, R. W.; Validation of numerical prediction for heat-transfer with airflow and frosting in fibrous insulation, paper 94-WA/HT-10. In: "Proc. ASME Int. Mech. Eng. Cong., Chicago." November **1994**.
74. MOD Specification UK/SC/4778A SCRDE. Moisture Vapour Transmission Test Method.
75. Morton, W. E.; and Hearle, J. W. S.; "Physical Properties of Textile Fibers." 3rd Ed. Manchester, UK: Textile Institute, **1993**.
76. Moyene, C.; Batsale, J. C.; and Degiovanni, A.; Approche experimentale et theorique de la conductivite thermique des milieux poreux humides, II: theorie. *Int. J. Heat Mass Transfer.* **1988**, *31(11)*, 2319–2330.
77. Mujumdar, A. S.; Handbook of Industrial Drying. New York: Marcel Decker; **1985**.

78. Nasrallah, S. B.; and Pere, P.; Detailed study of a model of heat and mass transfer during convective drying of porous media. *Int. J. Heat Mass Transfer.* **1988,** *31(5),* 957–967.
79. Nossar, M. S.; Chaikin, M.; and Datyner, A.; High intensity drying of textile fibers, Part I: the nature of the flow of air through beds of drying fibers. *J. Text. Inst.* **1973,** *64,* 594–600.
80. Patankar, S. V.; "Numerical Heat-transfer and Fluid Flow." New York: Hemisphere Publishing; **1980.**
81. Penner, S.; and Robertson, A.; Flow through fabric-like structures. *Text. Res. J.* **1951,** *21,* 775–788.
82. Provornyi, S.; and Slobodov, E.; Hydrodynamics of a porous medium with intricate geometry. *Theoret. Foundat. Chem. Eng.* **1995,** *29(1),* 1–5.
83. Rainard, L. W., Air permeability of fabrics I. *Text. Res. J.* **1946,** *16,* 473–480.
84. Rainard, L. W.; Air permeability of fabrics II. *Text. Res. J.* **1947,** *17,* 167–170.
85. Renbourne, E. T.; "Physiology and Hygiene of Materials and Clothing." Herts, England: Merrow Publishing Company; **1971.**
86. Saltiel, C.; and Datta, A. S.; Heat and mass transfer in microwave processing. *Adv. Heat Mass Transfer.* **1999,** *33,* 1–94.
87. Sanga, E.; Mujumdar, A. S.; and Raghavan, G. S.; Heat and mass transfer in non-homogeneous materials under microwave field. Presented at The 50th Canadian Chemical Engineering Conference. Montreal, Canada; October 15–18, **2000.**
88. Simacek, P.; and Advani, S.; Permeability model for a woven fabric. *Polym. Compos.* **1996,** *17(6),* 887–899.
89. Spencer-Smith, J. L.; The physical basis of clothing comfort, Part 3: water vapour transfer through dry clothing assemblies. *Cloth. Res. J.* **1977,** *5,* 82–100.
90. Spilman, L.; and Goren, S.; Model for predicting pressure drop and filtration efficiency in fibrous media. *Environ. Sci. Technol.* **1968,** *2,* 279–287.
91. Stanish, M. A.; Schajer, G. S.; and Kayihan, F.; A mathematical model of drying for hygroscopic porous media. *AICHE J.* **1986,** *32(8),* 1301–1311.
92. Toei, R.; Drying mechanism of capillary porous bodies. *Adv. Drying.* **1983,** *2,* 269–297.
93. Umbach, K. H.; Investigation of constructional principles for clothing textiles made of synthetic fibers worn next to the skin with good comfort properties. In: Technical Report No. AiF 3653. Boennigheim, Germany: Hohenstein Institute of Clothing Physiology; **1977.**
94. Umbach, K. H.; Moisture transport and wear comfort in microfiber fabrics. *Melliand Engl.* **1993,** *74,* E78–E80.
95. Von Hippel, A. R.; "Dielectric Materials and Applications." Boston: MIT Press; **1954.**
96. Waananen, K. M.; Litchfield, J. B.; and Okos, M. R.; Classification of drying models foe porous solids. *Drying Technol.* **1993,** *11(1),* 1–40.
97. Watkins, D. A.; and Slater, K.; The moisture vapour permeability of textile fabrics. *J. Text. Inst.* **1981,** *72,* 11–18.
98. Williams, A.; "Industrial Drying." London, England: Gardner/Leonard Hill Publishers; **1971.**

CHAPTER 20

A RESEARCH NOTE ON SORPTION-ACTIVE CARBON-POLYMER NANOCOMPOSITES

MARINA BAZUNOVA, RUSTAMTUKHVATULLIN, DENIS VALIEV, ELENA KULISH, and GENNADIJZAIKOV

CONTENTS

20.1 INTRODUCTION

Currently, the creation of materials with high adsorption activity to a range of definite substances by controlling their surface structure has significant interest. Particularly, selective sorbents for separation processes, dividing, or concentration of the components of different nature mixtures are developed on the basis of such composites.

The nature of the binder and the active components, and molding conditions are especially important at the process of sorption-active composites creating. These factors ultimately exert influence on the development of the porous structure of the sorbent particles and its performance. In this regard, it is promising to use powders of various functional materials having nanoscale particle sizes at the process of such composites creating. First, the high dispersibility of the particles allows them to provide a regular distribution in the matrix, whereby it is possible to achieve improved physical and mechanical properties. Secondly, high degree of homogenization of the components facilitates their treatment process. Third, it is possible to create the composites with necessary magnetic, sorption, dielectric, and other special properties combining volumetric content of components [1].

Powders of low-density polyethylene (LDPE) prepared by high-temperature shearing (HTS) used as one of prospective components of the developing functional composite materials [2, 3].

Development of the preparation process and study of physicochemical and mechanical properties of sorbents based on powder mixtures of LDPE, cellulose (CS), and carbon materials are conducted. As the basic sorbent material new-ultrafine nanocarbon (NC) obtained by the oxidative condensation of methane at a treatment time of 50 min (NC1) and 40 min (NC2) having a specific surface area of 200 m^2/g and a particle size of 30–50 nm is selected [4]. Highly dispersed form of NC may give rise to technological difficulties, for example, during regeneration of NC after using in gaseous environments, as well as during effective separation of the filtrate from the carbon dust particles. This imposes restrictions on the using of NC as an independent sorbent. In this connection, it should be included in a material that has a high porosity. LDPE and CS powders have great interest for the production of such material. It is known that a mixture of LDPE and CS powders have certain absorption properties, particularly, they were tested as sorbents for purification of water surface from petroleum and other hydrocarbons [5].

Thus, the choice of developing sorbents components is explained by the following reasons:

(i) LDPE has a low softening point, allowing conducting blanks molding at low temperatures. The small size of the LDPE particles (60–150 nm) ensures regular distribution of the binder in the matrix. It is also important that the presence of binder in the composition is necessary for maintaining of the material's shape, size, and mechanical strength.

(ii) Usage of cellulose in the composite material is determined by features of its chemical structure and properties. CS has developed capillary-porous structure that is why it has well-known sorption properties [5] toward polar liquids, gases, and vapors.

(iii) Ultrafine carbon components (nanocarbon, activated carbon (AC)) are used as functionalizing addends due to their high specific surface area.

20.2   EXPERIMENTAL

Fine powders of LDPE, CS, and a mixture of LDPE/CS are obtained by high-temperature shearing under simultaneous impact of high pressure and shear deformation in an extrusion type apparatus with a screw diameter of 32 mm [3].

Initial press-powders obtained by two ways. The first method is based on the mechanical mixing of ready LDPE, CS, and carbon materials' powders. The second method is based on a preliminary high-shear joint grinding of LDPE pellets and sawdust in a specific ratio and mixing the resulting powder with the powdered activated carbon (БАУ-Amark) and the nanocarbon after it.

Composites molding held by thermobaric compression at the pressure of 127 kPa.

Water absorption coefficient of polymeric carbon sorbents is defined by the formula: wherein m absorbed water is mass of the water, retained by the sorbent sample, m_{sample} is mass of the sample.

The adsorption capacity (A) of the samples under static conditions for condensed water vapor, benzene, n-heptane determined by method of complete saturation of the sorbent by adsorbate vapor in standard conditions at 20°C [6] and calculated by the formula: $A = m/(M \cdot d)$, wherein m—mass of the adsorbed benzene (acetone, n-heptane), g; M—mass of the dried sample, g; d—density of the adsorbate $\mathbf{6}$, g/cm^3.

Measuring of the tablets strength was carried out on the automatic catalysts strength measurer ПК-1.

Experimental error does not exceed 5 per cent in all weight methods at $P = 0.95$ and the number of repeated experiments $n = 3$.

20.3   RESULTS AND DISCUSSION

20.3.1   LDPE (OBTAINED BY THE METHOD OF HTS) POWDER PARTICLES' SIZE, DISPERSITY, AND SURFACE PROPERTIES STUDY

Powder components are used as raw materials for functional composite molding (including the binder LDPE), because molding of melt polymer mixtures with the active components has significant disadvantages. For example, the melt at high degrees of filling loses its fluidity, at low degrees of filling flow rate is maintained, but it is impossible to achieve the required material functionalization.

It is known that amorphous-crystalline polymers, which are typical heterogeneous systems, well exposed to high-temperature shear grinding process. For example, the process of HTS of LDPE almost always achieves significant result [3]. Disperse composition is the most important feature of powders, obtained as result of high-temperature shear milling. Previously, on the basis of the conventional microscopic measurement, it was believed that sizes of LDPE powder particles obtained by HTS are within 6–30 μm. Electron microscopy (Figure 20.1) gives the sizes of 60–150 nm. The active powder has a fairly high specific surface area (up to 2.2 m^2/g).

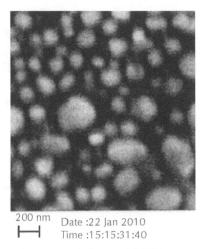

200 nm   Date :22 Jan 2010
⊢—⊣   Time :15:15:31:40

FIGURE 20.1   Electron microscopy of dispersed LDPE

The results of measurement of the water absorption coefficient and of the static capacitance of LDPE powder by n-heptane vapor are equal to 12 per cent and 0.26 cm³/g, respectively. Therefore, the surface properties of LDPE powder more developed than the other polyethylene materials'.

20.3.2 SELECTION OF MOLDING CONDITIONS OF SORBENTS BASED ON MIXTURES OF LDPE, CS, AND ULTRAFINE CARBON MATERIALS' POWDERS

Initial press-powders obtained by two ways. The first method is based on the mechanical mixing of ready LDPE, CS, and carbon materials' powders. The second method is based on a preliminary high-shear joint grinding of LDPE pellets and sawdust in a specific ratio and mixing the resulting powder with the powdered activated carbon and the nanocarbon after it. The method of molding—thermobaric pressing at a pressure of 127 kPa.

The mixture of LDPE/CS compacted into tablets at a temperature of 115–145°C was used as a model mixture for selection of composites molding conditions. Pressing temperature should be such that the LDPE softens but not melts, and at the same time forms a matrix to prevent loss of specific surface area in the ready molded sorbent due to fusion of pores with the binder. Data on the dependence of the coefficient of water absorption from temperature of the mixture of LDPE/CS (20/80 wt. %) pressing is presented in Table (20.1). It makes possible to determine the LDPE softening-melting temperature limits. The composites molded at a higher temperature, have a lower coefficient of water absorption than the tablets produced at a lower temperature that is why the lowest pressing temperature (120°C) is selected. At a higher content of LDPE, the water absorption coefficient markedly decreases with temperature.

TABLE 20.1　Dependence of water absorption coefficient (K) of the composites based on a mixture of LDPE/CS composition (20/80 wt. %) fromthepressingtemperature.

t, °C	115	119	125	128	133	136	138	142	145
K,%	180	180	182	177	168	169	165	150	149

Cellulose has a high degree of swelling in water (450%) [5], this may lead to the destruction of the tablets. Its contents in samples of composites, as it has been observed by the sorption of water, should not exceed 30 wt. %. There is a slight change of geometric dimensions of the tablets in aqueous medium at an optimal value of the water absorption coefficient when the LDPE content is 20 wt. %.

Samples of LDPE/CS with AC, which sorption properties are well studied, are tested for selecting of optimal content of ultrafine carbon. As follows from Table (20.2), the samples containing more than 50 wt. % of AC have less water absorption coefficient values. Therefore, the total content of ultrafine carbon materials in all samples must be equal to 50 wt. %.

TABLE 20.2　Water absorption coefficient of PE/AC/CS composites samples

Composition of PE/AC/CS sorbent (wt. %)	20/5/75	20/15/65	20/25/55	20/35/45	20/45/35	20/50/30	20/65/15
K, %	136.3	136.5	135.9	138.1	135.6	133.7	120.7

Static capacitance measurement of samples, obtained from mechanical mixtures of powders of PE, CS, and AC, conducted on vapors of n-heptane and benzene, to determine the effect of the polymer matrix on the sorption properties of functionalizing additives. As can be seen (Table 20.3), with a decrease of the content of AC in the samples with a fixed (20 wt. %)

amount of the binder, reduction of vapor sorption occurs. It indicates that the AC does not lose its adsorption activity in the composition of investigated sorbents.

TABLE 20.3 Static capacitance of sorbents, A (cm³/g), by benzene and n-heptane vapors (20°C)

Composition of PE/AC/CS sorbent (wt. %)		20/0/75	20/15/65	20/35/55	20/45/35	20/60/20	20/75/15	20/80/0
A cm³/g	Byn-heptane vapors	0.08	0.11	0.16	0.19	0.23	0.25	0.25
	Byben-zene vapors	0.06	0.08	0.17	0.18	0.23	0.25	0.26

Strength of samples of sorbents (Figure 20.2) is in the range of 620–750 N. The value of strength is achieved in the following molding conditions: $t = 120°C$ and a pressure of 127 kPa.

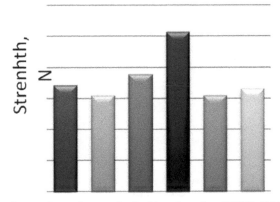

FIGURE 20.2 Comparison of strength of tablets, based on LDPE, CS (different species of wood) and AC powders. 1—sorbent of LDPE/AC/CS = 20/50/30 wt. % based on the powders of jointly dispersed pellets of LDPE and softwood sawdust with subsequently addition of AC; 2—sorbent of LDPE/AC/CS = 20/50/30 wt. % based on the powders of jointly dispersed pellets of LDPE and hardwood sawdust with subsequently addition of AC; 3—sorbent of LDPE/AC/CS = 20/50/30 wt. % based on the mechanical mixtures of the individual powders of LDPE, CS from softwood and AC; 4—AC tablet; 5—sorbent of LDPE/CS = 20/80 wt. %; 6—sorbent of LDPE/AC = 20/80 wt. %

Thus, optimal weight composition of the matrix of LDPE/CS composition—20/30 wt. % with 50 wt. % containing of carbon materials.

20.3.3 SORPTION PROPERTIES OF CARBON—POLYMER COMPOSITES BY CONDENSED VAPORS OF VOLATILE LIQUIDS

For a number of samples of sorbents static capacitance values by benzene vapor is identified (Figure 20.3). They indicate that the molded mechanical mixture of 20/25/25/30wt.% LDPE/AC/NC1/CS has a maximum adsorption capacity that greatly exceeds the capacity of activated carbon. High sorption capacity values bybenzene vapor appear to be determined by weak specific interaction of π-electron system of the aromatic ring with carbocyclic carbon skeleton of the nanocarbon [7].

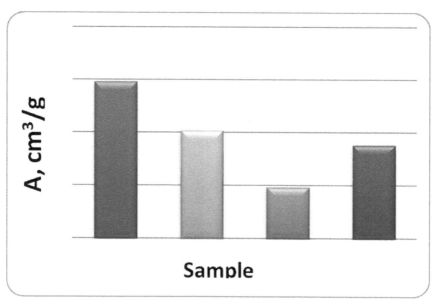

FIGURE 20.3 Static capacitance of sorbents, A (cm³/g) by benzene vapor (20°C).1—molded mechanical mixture of LDPE/AC/NC1/CS= 20/25/25/30 wt. %; 2—molded mechanical mixture of LDPE/AC/NC2/CS = 20/25/25/30 wt. %;3—molded mechanical mixture of LDPE/AC/CS=20/50/30 wt. %; 4—AC medical tablet (controlling)

Static capacitance of obtained sorbents by heptane vapors significantly inferiors to capacity of activated carbon (Figure 20.4), probably it is determined by the low polarizability of the molecules of low-molecular alkanes. Consequently, the investigated composites selectively absorb benzene and can be used for separation and purification of mixtures of hydrocarbons.

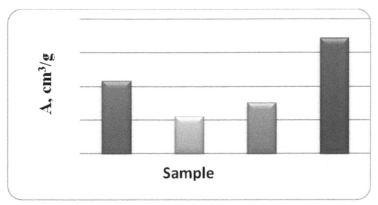

FIGURE 20.4   Static capacitance of sorbents, A (cm^3/g) byn-heptane vapor (20°C). 1—molded mechanical mixture of LDPE/AC/NC1/CS= 20/25/25/30 wt. %; 2—molded mechanical mixture of LDPE/AC/NC2/CS = 20/25/25/30 wt. %; 3—molded mechanical mixture of PE/AC/CS=20/50/30 wt. %; 4—AC medical tablet (controlling)

Molded composite based on a mechanical mixture of PENP/AC/NC1/CS = 20/25/25/30 wt. % has a sorption capacity by acetone vapor comparable with the capacity of activated carbon (0.36 cm^3/g) (Figure 20.5).

Sorbents' samples containing NC2 have a low value of static capacity by benzene, heptanes and acetone vapor. It can be probably associated with partial occlusion of carbon material pores by remnants of resinous substances-byproducts of oxidative condensation of methane, and insufficiently formed porous structure.

The residual benzene content measuring data (Table 20.4) shows that the minimal residual benzene content after its desorption from the pores at t = 70°C for 120 minob serves in case of sorbent LDPE/AC/NC1/CS composition = 20/25/25/30 wt. %. It allows to conclude that developed sorbents have better ability to regenerate under these conditions in comparison with activated carbon.

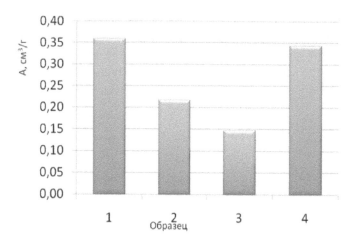

FIGURE 20.5   Static capacitance of sorbents, A (cm³/g) acetone vapor (20°C). 1—molded mechanical mixture of LDPE/AC/NC1/CS= 20/25/25/30wt. %; 2—molded mechanical mixture of LDPE/AC/NC2/CS = 20/25/25/30 wt. %; 3—molded mechanical mixture of LDPE/AC/CS=20/50/30 wt. %; 4—AC medical tablet (controlling)

TABLE 20.4   Sorbents' characteristics: total pore volume V_{tot}; static capacitance (A) by benzene vapors at the sorption time of 2 days; residual wt. fracrion of the absorbed benzene after drying at t=70°C for 120 min

LDPE/AC/NC/CS sorbent composition (wt. %)	$V_{tot.}$ (cm³/g)	A(cm³/g)	Residual benzene content as a result of desorption (%)
20/25/25/30	1.54	0.5914	2.9
20/50/-/30	1.21	0.1921	10.3
-/100/-/-	1.6	0.3523	32.0

Thus, the usage of nanosized LDPE as a binder gives a possibility to get the molded composite materials with acceptable absorption properties. Varying the ratio of the components of the compositions on the basis of ternary and quaternary mixtures of powdered LDPE, cellulose, and ultra-fine carbon materials it is possible to achieve the selectivity of sorption properties by vapors of certain volatile liquids. These facts allow to suggest that the proposed composites are expedient to use for separation and purification of gaseous and steam mixtures of different nature.

Developed production method of molded sorption-active composites based on ternary and quaternary mixtures of powdered LDPE, cellulose, and ultrafine carbon materials can be easily designed by equipment and can be used for industrial production without significant changes.

20.4   CONCLUSIONS

(i)   It is revealed by electron microscopy that the dispersed LDPE particle sizes are 60–150 nm. It allows to obtain functional composite materials with a regular distribution of components and with the necessary physicochemical properties, depending on their volume content, on its base.

(ii) Optimal conditions for molding of sorbents on the basis of three- and four-component mixtures of powdered LDPE, cellulose, and ultrafine carbon materials were determined: temperature 120°C and pressure of 127 kPa, content of the binder (LDPE) = 20 wt. %.

(iii) Established that molded mechanical mixture of LDPE/AC/NC1/CS (20/25/25/30wt. %) has a static capacity (by condensed vapors of benzene and acetone)= 0.6 and 0.36 cm^3/g respectively, what exceeds the capacity of activated carbon. The static capacitance of the compositions by the n-heptane vapors is 0.21 cm^3/g, therefore, the proposed composites are useful for separation and purification of gaseous and steam mixtures of different nature.

KEYWORDS

- **Cellulose**
- **High-temperature shift crushing**
- **Nano-carbon**
- **Polyethylene**
- **Sorbents**

REFERENCES

1. Akbasheva, E. F.; and Bazunova, M. V.; Tableted sorbents based on cellulose powder mixtures, polyethylene and ultra-dispersed carbon. In: Materials Open School Confer-

ence of the CIS "Ultra-fine and Nanostructured Materials." October 11–15, 2010, Ufa: Bashkir State University, **2010,** 106 p.

2. Enikolopyan, N. S.; Fridman, M. L.; and Karmilov, A. Yu.; Elastic-deformation grinding of thermo-plastic polymers. *Rep. USSR.* 1987, *296(1),* 134–138.

3. Akhmetkhanov, R. M.; Minsker, K. S.; and Zaikov, G. E.; On the Mechanism of Fine Dispersion of Polymer Products at Elastic Deformation Effects. Plasticheskie Massi; 2006, *8,* 6–9.

4. Aleskovskiy, V. B.; and Galitseisky, K. B.; Russian Patent "Method of Ultrafine Carbon" No 2287543 from November 20, **2006.**

5. Raspopov, L. N.; Russiyan, L. N.; and Zlobinsky, Y. I.; Water proof composites comprising wood and polyethylene dispersion. *Russ. Polym. Sci. J.* **2007,** *50(3),* 547–552.

6. Keltsev, N. V.; Fundamentals of Adsorption Technology. Moscow: Chemistry; **1984,** 595 p.

7. Valinurova, E. R.; Kadyrov, A. D.; and Kudasheva, F. H.; Adsorption properties of carbonrayon. *Vestn. Bashkirs. Univer.* **2008,** *13(4),* 907–910.

INDEX